Arts, Research, Innovation and Society

Creativity in general and the arts in particular are increasingly recognized as drivers of cultural, economic, political, social, and scientific innovation and development. In Art and Research, some of the principal questions to be explored by the ARIS project, are outlined: • Could and should Artists be Researchers? • How are the systems of the Arts and the Sciences connected and /or disconnected? • What is the position and status of the Arts in defining the terms "progress" and "development"? The Springer ARIS series explores - at the macro, meso and micro levels and in terms of qualitative as well as quantitative studies - theories, policies and practices about the contributions of artistic research and innovations towards defining new forms of knowledge, knowledge production as well as knowledge diffusion, absorption and use. Artistic research, artistic innovations and arts-based innovations have been major transformers as well as disruptors of the ways in which societies, economies, and political systems perform. Ramifications here refer to the epistemic socio-economic, socio-political and socio-technical base and aesthetic considerations on the one hand, as well as to strategies, policies, and practices on the other, including sustainable enterprise excellence considerations in the context of knowledge economies, societies and democracies. The series features research monographs, edited volumes, proceedings, Briefs, and textbooks, and may also include handbooks and reference works, and in-print as well on-line rich media encapsulations of ideas and insights, representing cutting-edge research and the synthesis of a body of work in the field.

Dennis Del Favero • Susanne Thurow
Michael J. Ostwald • Ursula Frohne
Editors

Climate Disaster Preparedness

Reimagining Extreme Events through Art and Technology

Springer

Editors
Dennis Del Favero
iCinema Centre for Interactive Cinema Research
The University of New South Wales
Kensington, NSW, Australia

Susanne Thurow
iCinema Centre for Interactive Cinema Research
The University of New South Wales
Kensington, NSW, Australia

Michael J. Ostwald
School of Built Environment
The University of New South Wales
Kensington, NSW, Australia

Ursula Frohne
Centre for Advanced Study
Universität Münster
Münster, Nordrhein-Westfalen, Germany

ISSN 2626-7683 ISSN 2626-7691 (electronic)
Arts, Research, Innovation and Society
ISBN 978-3-031-56113-9 ISBN 978-3-031-56114-6 (eBook)
https://doi.org/10.1007/978-3-031-56114-6

This Springer imprint is published by the registered company Springer Nature Switzerland AG
The registered company address is: Gewerbestrasse 11, 6330 Cham, Switzerland

Paper in this product is recyclable.

Foreword

It is well established that climate change, caused by the burning of coal, oil and gas, has supercharged extreme weather events, including heatwaves, fires, floods and storms. In Australia, the unprecedented Black Summer bushfires of 2019–20 devastated communities across the country and were then followed by three years of record floods. In 2023, the Northern Hemisphere was hit by record heatwaves and wildfires, while massive floods killed thousands.

In the past, Australians would have expected catastrophic events to happen irregularly, with long peaceful periods in between. Today, extreme, compounding events are occurring with a constancy that leaves little time for recovery to communities or emergency responders. Our capacity to prepare and respond has also been substantially compromised. Windows for fuel reduction burning have dramatically shortened, and overlapping fire seasons between states and countries have reduced our ability to share personnel and major equipment, such as aerial water bombers.

As the risk of climate-fuelled disasters worsens, governments, communities and even emergency services are struggling to comprehend the nature of the rapidly changing threat environment. Most scientific simulations that assist in modelling fires and floods usually show predictions on maps. This means that communities, unable to easily access or interpret technical models, lack a framework to preview and visualise unpredictable dynamics and impact of extreme fires and floods in their locality. Therefore, they struggle to realistically rehearse a suitable response. History is no longer a reliable indicator of what to expect, as shown when flooding in the northern NSW town of Lismore peaked more than two metres over the previous historic high, and during Black Summer, when the number of homes destroyed by fires in New South Wales exceeded previous records by more than 11 times.

The ability to visualise and viscerally understand what disasters look and feel like is critical in helping people increase preparedness, so there is an urgent need to work with emergency services, industry, researchers and community members to develop new capabilities using latest advances in the arts, AI and sciences. Tools that enable communities to previsualise threats; see how they are likely to unfold; then, acting on this knowledge, rehearse effective response strategies tailored to their particular circumstances will become more and more essential as risks intensify and change.

In September 2023, the Australian Bureau of Meteorology officially declared the Pacific to be swinging from three years of La Niña into El Niño, as a positive Indian Ocean Dipole developed to Australia's northwest. This puts Australia squarely on track for prolonged dry and intensely hot weather patterns that could put coming bushfire seasons on steroids. A rare triple La Niña event drenched soils and supported prolific forest and grass growth, as well as regrowth in areas hit by Black Summer. While this provided much-needed reprieve for both the land and fire-affected communities, the increased fuel loads have rapidly dried out in much of Australia during the 2022 winter, the warmest winter on record.

This year's scorching Northern Hemisphere summer paints a sobering picture confirming the most-dire predictions of climate scientists. More sobering still, companies, industries and nations continue to burn massive amounts of coal, oil and gas. Extreme summer heatwaves affected the United States, Africa, China, Japan and Europe, with Canada suffering through its own version of Black Summer. It is important to note that the experience of the Northern Hemisphere was not exacerbated by an El Niño, which is a Southern Hemisphere phenomenon. Words such as 'unprecedented' in combination with these disasters start to ring hollow as record after record is toppled in the wake of ever greater extremes. Anyone tempted to feel a degree of assurance of a peaceful summer should remember that it only took three weeks of extremely dry conditions to set the scene for the devastating Maui wildfire in August that killed at least 100 people and wiped out the town of Lahaina—becoming 'the worst natural disaster in the history of Hawaii'. Another record that led to immense ecological and human suffering.

As Australia entered into spring, 'Extreme' and 'Catastrophic' fire danger warnings were issued in the Northern Territory, Western Australia, Queensland, New South Wales and Victoria. Western Australia declared its fire season two months early, with a blaze taking hold in November at the urban/bushland interface north of Perth growing to over 1,500 hectares overnight, destroying 18 homes. In Queensland, dry thunderstorms sparked hundreds of bushfires that claimed five lives and 65 homes before the end of October. By then, a singe fire in the Northern Territory had already burnt over one million hectares. The consensus among scientists and fire services is that despite some November rains in the east, the 2023–24 fire season is likely to be extensive and volatile. Yet, the coming years are of chief concern, due to increasingly unpredictable dynamics and the likelihood of a return to drought.

During an El Niño phase, unpredictability is what keeps responders and scientists awake at night. With seasonal rhythms thrown into disarray, on-the-ground conditions can change rapidly, from moist and cool to dry and hot, challenging our predictive models and firefighting strategies. In early December, after weeks of storms and rains, most of Australia was affected by severe heatwaves and the return of 'Catastrophic' and 'Extreme' fire dangers.

Climate change reinforces such effects. For example, on the NSW South Coast, winter rainfall has reduced by 12% over the last 20 years, creating drier forests and grasslands, and in the south-west of Western Australia, by 20%. Drier fuels lead to more violent and erratic fire behaviour, including pyro-convective (i.e. fire-generated) storms or fire tornados. Fires burning in hot, dry conditions can create extreme heat that generates local weather systems, giving rise to towering pyrocumulonimbus clouds that trigger intense wind downbursts and dry lightning and drive fire fronts at frightening speeds. Between 1978 and 2018, there were 60 recorded pyro-convective storms across Australia. In a six-month period in 2019–20, there were a staggering 29 pyro-convective events.

Another worrying change is that fires now often burn as intensely through the night as they do during the day in periods of extreme heat and low humidity, giving little relief and robbing firefighters of opportunities to put in place previously reliable containment strategies. For decades, our chief mitigation strategy has been to conduct hazard-reduction burns, i.e. lighting controlled fires during cooler weather to reduce fuel levels that could feed fires in hotter months. Climate change is increasingly reducing time windows of safe weather to conduct these burns—during La Niña it was too wet to burn, and in spring 2023 it rapidly became too hot, windy and dangerous to burn safely. After decades of confronting and battling devastating fires, we know these observations to be indicators of a dangerously transforming climate, something that fire historian and author Stephen Pyne has termed the 'Pyrocene'—the era of fire. Increased greenhouse gas concentrations have made the weather wilder and more dangerous than humans have ever experienced before.

History only provides limited instruction on what to expect as we venture into the unchartered territory that our future holds. Among scientists, the abnormally high sea temperatures over recent months have triggered a red alert, warning that we are speeding towards dangerous terrain of impending compound crises and climate tipping points. Emergency services and those communities who have already been at the frontline of ever worsening floods and fires have seen their devastation, smelt their acrid stench and felt the terror in their bones. Many will grapple with the horrors of these events for a lifetime.

Unfortunately, we know that even if the world starts to rapidly reduce greenhouse gas pollution, we are progressing into a warmer and more volatile world. The shape of our world in decades and centuries to come is being determined by our actions today. On the current trajectory, temperatures will keep going up. Heatwaves, fires, storms and floods will continue to worsen. The solution is to rapidly reduce the pollution that is driving the problem—the burning of coal, oil and gas. The goal right now must be to stabilise and eventually lower global temperatures. We now have an Australian government that accepts the science of climate change and is progressing some effective policies. However, fossil-fuel companies are still attempting to open more polluting projects and are highly resistant to change. In the 2023/24 financial year alone, projected fossil-fuel industry subsidies outweigh the federal Disaster Relief Fund at a ratio of 55:1. We urgently need to divert this money from polluting industries to communities across Australia so that we can more effectively safeguard them against heatwaves, fires, floods, cyclones and droughts.

As climate change is here with us today, we must both tackle the root cause, fossil fuels, and prepare for the consequences we cannot avoid. A fundamental question is how we can be better prepared to address extreme fires and floods. Due to their unforeseen and escalating dynamics, communities need to have the ability to simulate events in advance, previewing and actually seeing how they could evolve in their specific geographic locations, then rehearse what they will do in response. In short, communities need to be able to deeply envision what could come their way so that they have a head start in getting ready.

The research presented in this book offers a compelling vision of how we might help to build such preparedness. It draws together advances from disciplines as varied as artificial intelligence, creative arts and climate science to map cutting-edge capabilities that can pilot novel approaches to visualising and rehearsing plausible and probable extreme climate scenarios. This is crucial as we prepare Australian communities for an increasingly challenging future.

Climate Council of Australia/ELCA
Potts Point, NSW, Australia
Greg Mullins

Climate Council of Australia
Potts Point, NSW, Australia
Amanda McKenzie

Contents

Contributors

Gilbert Accary is an Associate Professor of fluid and thermal sciences at the Lebanese American University. He is specialised in thermo-fluid systems design and in computational fluid dynamics and conducts his main research activity in the field of wildfire modelling. He made a significant contribution to the development of a 3D physics-based tool for the prediction of wildfire behaviour and is co-author of two patents, in wildfire modelling and in fire safety engineering. Email: gilbert.accary@lau.edu.lb

Ali Asadipour leads the UK's Royal College of Arts (RCA)'s Computer Science Research Centre, specialising in immersive experiences and AI. His international influence is amplified by holding Senior Visiting Fellowship positions at the University of New South Wales and Korea Advanced Institute of Science & Technology, facilitating collaborations with esteemed global research labs and assuming pivotal roles in national knowledge exchange networks. Before joining RCA, he spearheaded interdisciplinary research groups at Warwick University, overseeing a multitude of internationally funded projects. With a PhD, MSc and BSc (Hons) in computer science and engineering, he possesses over a decade of academic and industrial acumen and has co-founded and held the role of Chief Technology Officer in technology-focused enterprises. Email: ali.asadipour@rca.ac.uk

Susan Atkinson is a senior strategic communication consultant and a PhD candidate in the News and Media Research Centre at the University of Canberra. Her research focus is on people's information needs and communication behaviours during disasters and emergencies and how community communication ecologies spontaneously form to facilitate the flow of information and help communities respond. Email: sue.atkinson@canberra.edu.au

Alethea Blackler is a Professor in design and Associate Dean (Research) of the Faculty of Creative Industries, Education, and Social Justice at Queensland University of Technology. Her research interests are in intuitive interaction, in particular in relation to older people and technology, design for dementia and novel technologies for active play for young children. She has worked extensively with government, community and industry partners and has led three large-scale Australian Research Council grants. She has published more than 100 peer-reviewed papers, been regularly invited to give presentations internationally and received several awards, including a 2018 Ig Nobel. Email: a.blackler@qut.edu.au

Baylee Brits is an Office of National Intelligence Postdoctoral Research Fellow at The University of New South Wales's iCinema Centre. Her work looks at storytelling, semiosis, text analysis and abduction in artificial intelligence. She holds a PhD from UNSW and an MA from the University of Amsterdam. She is the author of *Literary Infinities: Number and Narrative in Modern Fiction* (2017) and co-editor of *Aesthetics After Finitude* (2016) and *Covert Plants* (2018). Email: b.brits@unsw.edu.au

Carlos Tirado Cortes is a Post doctoral Research Fellow at The University of New South Wales's iCinema Centre. His current work focuses on how people interact with different virtual environments. Other research interests include virtual reality and its uses, interaction methodologies for mixed reality, immersive visualisation and the study of cybersickness. Email: c.tirado@unsw.edu.au

Seán Cubitt is Professor of screen studies at the University of Melbourne. His publications include *The Cinema Effect* (2004), *EcoMedia* (2005), *The Practice of Light* (2014), *Finite Media* (2017), *Anecdotal Evidence* (2020) and *Truth (Aesthetic Politics 1)* (2023). Co-editor of *Ecocinema 2* (2023), *The Ecocinema Reader: Theory and Practice* (2012) and *Ecomedia: Key Issues* (2015), he is series editor for Leonardo Books at MIT Press. He researches the history and philosophy of media, ecopolitical aesthetics, media arts and technologies and media art history. Email: sean.cubitt@unimelb.edu.au

Stuart Cunningham is Professor, Faculty of Art and Design, University of Canberra; Distinguished Emeritus Professor, Queensland University of Technology; and Senior Associate, Outside Opinion. He was Director of the first Australian Research Council Centre of Excellence based outside the hard sciences ('Creative Industries and Innovation') from 2005 to 2014. He is a leading researcher in the role that cultural and creative industries and enterprise play in contemporary social and economic innovation. Email: stuart.cunningham@canberra.edu.au

Jane W. Davidson Fellow of the Australian Academy for the Humanities, undertakes research in performance, musical development, intercultural engagement and music for wellbeing outcomes. Her roles have included: editor of *Psychology of Music* (1997–2001), president of both the Musicological Society of Australia (2010–11) and the Australian Music and Psychology Society (2018–21) and deputy director of the Australian Research Council Centre of Excellence for the History of Emotions (2011–18). She is currently Head of Performing Arts and Chair of the

University of Melbourne's Creativity and Wellbeing Initiative. Jane is an opera singer and director who has secured grants and awards in Australia and overseas. Email: j.davidson@unimelb.edu.au

Dennis Del Favero is an artist, ARC Australian Laureate Fellow and Chair Professor of Digital Innovation and Executive Director of The University of New South Wales's iCinema Centre. He has led numerous large-scale interdisciplinary art projects that explore the visceral dynamics of unpredictable climate scenarios and the aesthetics of uncertainty using AI visualisation systems. He is a Senior Visiting Fellow at Royal College of Art, UK; Editorial Board Member of Quodlibet Studio. Corpi, Italy; and former Executive Director of the Australian Research Council | Humanities & Creative Arts. He is represented by Galerie Brigitte Schenk, Cologne and Mais Wright, Sydney. Email: d.delfavero@unsw.edu.au

Marcus Foth is a Professor of urban informatics at Queensland University of Technology's School of Design and a Chief Investigator at its Digital Media Research Centre. For over two decades, he has led ubiquitous computing and interaction design research into interactive digital media, screen, mobile and smart-city applications. Marcus founded the Urban Informatics Research Lab in 2006 and the QUT Design Lab in 2016. He has published more than 280 peer-reviewed publications and is a Fellow of the Australian Computer Society and the Queensland Academy of Arts & Sciences, a Distinguished Member of the international Association for Computing Machinery (ACM) and currently serves on Australia's national College of Experts. Email: m.foth@qut.edu.au

Ursula Frohne is Professor for art history at the Universität Münster (Germany). Previously, she was Professor of twentieth and twenty-first-century art history at the University of Cologne, Chief Curator for ZKM Center for Art & Media in Karlsruhe and Lecturer at Karlsruhe's State Academy of Fine Art. In Cologne, she chaired the DFG-project Cinematographic Aesthetics in Contemporary Art (2007–14), while from 2023 she has been co-leading the DFG-Centre for Advanced Study: Access to Cultural Goods in Digital Change. She was awarded the University of Cologne's Leo Spitzer Prize for Excellence in Research. She has published widely on contemporary art practice and technological creative media. Email: ursula.frohne@uni-muenster.de

Janet Fulton is an independent researcher in communication and media, an Adjunct Associate Professor at the Royal Melbourne Institute of Technology and a Research Associate at the News and Media Research Centre at the University of Canberra. Fulton has worked on projects in journalism, journalism education, media entrepreneurship, creativity and cultural production, and diversity. Email: janet.fulton@canberra.edu.au

Charles Green is Professor of contemporary art at The University of Melbourne. He has written *The Third Hand: Artist Collaborations from Conceptualism to*

Postmodernism (University of Minnesota Press, 2001), and (with Anthony Gardner) *Biennials, Triennials, and Documenta* (Wiley-Blackwell, 2016). He works with long-time collaborator Lyndell Brown as one artist and they were Australia's Official War Artists in Iraq and Afghanistan in 2007–08. Their works have been acquired by most major Australian public art museums and private collections, including the National Gallery of Australia, the National Gallery of Victoria, the Art Gallery of New South Wales and the Australian War Memorial. They are represented by ARC One Gallery. Email: cdgreen@unimelb.edu.au

Helena Grehan is Vice Chancellor's Professorial Research Fellow at the Western Australian Academy of Performing Arts, based at Edith Cowan University. Previously, she was Professor of theatre, performance and creative arts at Murdoch University. She has published essays and books on spectatorship and ethics, intercultural theatre, performance and politics and new media dramaturgy. Her publications include *Performance, Ethics and Spectatorship in a Global Age* (2009), *New Media Dramaturgy* (with Peter Eckersall and Edward Scheer, 2017) and *The Routledge Companion to Theatre and Politics* (co-edited with Eckersall, 2019). Grehan is currently Lead CI on an ARC Linkage Grant to Digitise Western Australia's Vulnerable Cultural Heritage. She is Deputy Editor of the leading performance studies journal, *Performance Research*. Email: h.grehan@ecu.edu.au

Mick Grierson is Research Leader and Professor at University of the Arts London's Creative Computing Institute. His research explores new approaches to creating sounds, images, video and interactions through signal processing, machine learning and information-retrieval techniques. Hardware and software based on his research has been widely used by world-leading production companies, tech start-ups and artists, including the BBC, Massive Attack and many others. He led the UK's first BSc programs in creative computing and the world's first creative programming MOOC. Work by his students and research group has received over £6.4 million in funding and was instrumental in developing the field known as creative AI. Email: m.grierson@arts.ac.uk

Linda Hassall is a researcher, theatre maker and educator in drama, applied performance and performance studies. She explores the intersection between theatre and social justice themes derived from climate change, recently publishing *Theatres of Dust: Climate Gothic Analysis in Contemporary Australian Drama and Performance Landscapes* (2021). She is Field of Study Coordinator, Drama and Program Director, Creative Industries at Griffith University, Australia. Email: l.hassall@griffith.edu.au

Ahmad Hassan is a postgraduate research student at Victoria University. His current research primarily focuses on modelling extreme wildfire phenomena using physics-based simulation tools. He is also affiliated with the Natural Hazards Research Australia Centre, where his research is dedicated to advancing resilience and reducing wildfire disaster risk. Email: ahmad.hassan6@live.vu.edu.au

Nagida Helsby-Clark is a design researcher and PhD candidate at The University of New South Wales's iCinema Centre, supported by a scholarship from CSIRO's Data61—the data and digital specialist arm of Australia's national science agency. Her research focuses on immersive aesthetics and embodied sense making for climate extremes. She conducts experimental research across Data61 and iCinema Centre, in particular on the ARC-funded iFire Laureate project (led by Prof. Dennis Del Favero). Nagida holds a Master in Visualisation, Simulation and Immersive Design from UNSW (2021) and has a background as a community organiser and climate-justice advocate. Email: n.helsbyclark@student.unsw.edu.au

Fiona M. Johnson is the Director of The University of New South Wales's Water Research Centre and an Associate Professor in the School of Civil and Environmental Engineering. She has over 20 years' experience in hydrology, working as a consultant for government and in academia. Fiona's areas of research and teaching focus on statistical hydrology, particularly with respect to flooding and extreme events and the use of global climate models for climate change assessments of water resource systems. She has a particular interest in solutions to climate and hydrological challenges faced by communities in the Global South. Email: f.johnson@unsw.edu.au

Lara Mani is a Senior Research Associate at the Centre for the Study of Existential Risk at the University of Cambridge, where her research focuses on establishing effective risk communication strategies for global catastrophic risks, with a focus on volcanic risk. Her interdisciplinary research adopts novel and creative approaches to risk communication such as scenario-based exercises and role-playing games. Email: lm881@cam.ac.uk

Kerry McCallum is Director of the News and Media Research Centre at University of Canberra. Her research specialises in the relationships between changing media and Australian social policy. She is the co-author of *The Dynamics of News and Indigenous Policy in Australia* (Intellect, 2017) and is lead investigator on the ARC-funded project *Breaking Silences: Media and the Child Abuse Royal Commission*. Email: kerry.mccallum@canberra.edu.au

Jon McCormack is an artist and researcher in computing. He is Founder and Director of Monash University's SensiLab, where he holds a Professorship in computer science jointly with the faculties of Information Technology and Art Design and Architecture. His research interests include generative art, design and music, evolutionary systems, computer creativity, human–machine relationships and creative artificial intelligence. He won an ARC Future Fellowship, has led six ARC Discovery projects and participated on three Linkage projects. Jon holds a PhD in computer science and an Honours degree in applied mathematics and computer science, both from Monash University, as well as a Graduate Diploma of Art (Film and Television) from Swinburne University. Email: jon.mccormack@monash.edu

Amanda McKenzie is Co-Founder and CEO of the Climate Council of Australia. She is a prominent Australian environmental leader and one of the best-known public commentators on the climate crisis in Australia. Previously, she was the Senior Communications Advisor at the Climate Commission and Co-Founder of the Australian Youth Climate Coalition. She served on renewable energy expert panels for the Queensland and Northern Territory governments. Amanda has won numerous awards, including being recognised as one of Westpac's 100 Women of Influence, *Sunday Life* magazine's Top 25 Women Reshaping Australia and a finalist in the Not-for-Profit CEO of the Year Awards 2023. Email: amanda@climatecouncil.org.au

Khalid Moinuddin is a Professor of fire modelling at Victoria University. He is an internationally recognised expert in physical modelling of fire dynamics in building, wildland, tunnel and naval settings. He sits on the Victorian Building Appeal Board to ensure safety in residential and commercial buildings. Email: khalid.moinuddin@vu.edu.au

Greg Mullins is an internationally recognised expert in responding to major bushfires and natural disasters, with more than 50 years' experience as a firefighter. He coordinated responses to many major disasters over more than two decades and studied bushfire control in France, Spain, Canada and the United States. He retired as Commissioner of Fire and Rescue New South Wales in 2017, after 39 years of service and nearly 14 years in the role. On retirement, Greg immediately re-joined the rural fire brigade where he started as a volunteer in 1972 and fought fires throughout the state during the 2019–20 Black Summer. In early 2019, he formed Emergency Leaders for Climate Action (ELCA), a coalition of former chiefs and deputy chiefs from every fire service in Australia who tried to warn the Australian government of the pending bushfire catastrophe. He is a Climate Councillor and writer. Email: gregmullins@outlook.com.au

Michael J. Ostwald is a Professor of architectural analytics at The University of New South Wales. He was previously an EU Distinguished Professor, Dean of the Built Environment at the University of Newcastle, professorial fellow at Victoria University Wellington and Adjunct Professor at Royal Melbourne Institute of Technology. He completed postdoctoral research on applications of spatial computing in architecture at University of California LA, Canadian Centre for Architec ture (McGill, Montreal) and Harvard (Mass.). He is the author of 17 books and over 140 journal articles. Michael is Editor-in-Chief of the *Nexus Network Journal: Architecture and Mathematics* (Springer) and sits on the editorial boards of *Archite ctural Research Quarterly* (Cambridge) and *Architectural Theory Review* (Taylor & Francis). Email: m.ostwald@unsw.edu.au

Maurice Pagnucco received his BSc and PhD degrees in computer science from The University of Sydney in 1991 and 1996, respectively. He is currently a Professor and the Deputy Dean (Education) of the Faculty of Engineering at the University of New South Wales (UNSW). He joined UNSW in 2001 and has held the position of Deputy Dean (Education) since 2015. He was previously the Head of the School of

Computer Science and Engineering at UNSW from 2010 to 2019. His research is focused on artificial intelligence, with particular emphasis on knowledge representation and reasoning, and cognitive robotics. Email: morri@unsw.edu.au

Sora Park is a Professorial Research Fellow at the News and Media Research Centre at University of Canberra. Her research focuses on digital media users, media markets and media policy. She is the leader of the Digital News Report: Australia project and the Lead Investigator of the ARC Discovery Project *The Rise of Mistrust: Digital Platforms and Trust in News Media*. Email: sora.park@canberra.edu.au

Simonne Pengelly holds a BA(Hons), MBA, and GCertDRR and is a PhD candidate in disaster management at the University of Newcastle. She conducts research in disaster risk reduction, with a focus on expanded roles for innovation and arts-based creativity. Her research concentrates on the pre-disaster (preparedness) phase, examining pathways to resilient social transformation via transdisciplinary investigation. Simonne is also a recipient of the prestigious Vice Chancellor's Academic Scholarship. She has a background in film, television and drama and as a touring and recording musician. More recently, she has worked in politics, business and law with a focus on strategic future planning. Email: simonne.pengelly@uon.edu.au

Ashish Sharma is a Professor of hydrology and water resources at the School of Civil and Environmental Engineering at The University of New South Wales. He has authored over 300 journal publications covering various facets of the discipline, including flood design under the impacts of climate change and graduated nearly 40 PhDs, many of whom are senior academics in different parts of the world. Ashish has served as the President of the International Association of Hydrologic Sciences (IAHS), the Commission on Statistical Hydrology (STAHY) and currently serves as the Editor-in-Chief of the *Journal of Hydrology-X*. Email: a.sharma@unsw.edu.au

Belinda Smaill is Professor of film and screen studies at Monash University in Melbourne. Her work focuses on nonfiction and documentary screen culture. She is currently researching the ethical, cultural and institutional issues that pertain to the presentation of the environment and biodiversity on screen. She is the author of *The Documentary: Politics, Emotion, Culture* (2010) and *Regarding Lif*Email: *Animals and the Documentary Moving Image* (2016) and co-author of *Transnational Australian Cinema: Ethics in the Asian Diasporas* (2013). She has published widely in international journals, including *Screen*, *Journal of Cinema and Media Studies* and *Camera Obscura*. Email: belinda.smaill@monash.edu

Yang Song received her BEng in computer engineering from Nanyang Technological University, Singapore, in 2003 and PhD in Computer Science from the University of Sydney in 2013. She is currently a Scientia Associate Professor and ARC Future Fellow in the School of Computer Science and Engineering at The University of New South Wales, Australia. She was an ARC DECRA Fellow at The University of Sydney from 2015 to 2018. Her research interests include computer vision, biomedical image analysis and machine/deep learning. Email: yang.song1@unsw.edu.au

Susanne Thurow is a Deputy Director and ARC Laureate Postdoctoral Research Fellow at The University of New South Wales's iCinema Centre, where she leads the climate aesthetics research program. Her interdisciplinary work rethinks contemporary arts through performative digital aesthetics, having co-developed multidisciplinary projects with industry partners, such as Opera Australia. She has published widely, spanning theatre, performance and digital media studies. Her latest book *Performing Indigenous Identities on the Contemporary Australian Stage* (Routledge, 2020) won the 2021 Alvie Egan Award and the 2019 UNSW Art & Design Dean's Award for Research Excellence, Best Monograph. In the past, she worked for Thalia Theater (Germany), Big *h*ART (Australia) and German cultural association Goethe-Institut. Email: s.thurow@unsw.edu.au

S. Travis Waller is the Lighthouse Professor and Chair of Transport Modelling and Simulation at the Technical University of Dresden, Germany, as well as a Professor at the Australian National University (ANU). He was previously a Professor and Deputy Dean (Research) for the Faculty of Engineering at UNSW, where he was the Founding Director of the Research Centre for Integrated Transport Innovation (rCITI). Waller is a global research leader in the domain of transportation network modelling and simulation. He has secured more than $30 million in research funding, has published more than 200 peer-reviewed journal papers and supervised 38 PhD students to completion. Email: steven_travis.waller@tu-dresden.de

Gabriele Wong-Parodi is an Assistant Professor in the Department of Earth System Science and Social Sciences Division at the Stanford Doerr School of Sustainability and a Center Fellow at the Woods Institute for the Environment at Stanford University. Her research focuses on understanding human responses to global environmental change and developing and testing robust interventions to support informed decision making to address that change. Email: gwongpar@stanford.edu

Sarah Woodland is a researcher, practitioner and educator in applied theatre, participatory arts and socially engaged performance. She has recently completed a three-year Dean's Fellowship in the Faculty of Fine Arts and Music, The University of Melbourne, investigating how the performing arts can promote social justice and wellbeing in institutions and communities. Sarah is a Senior Lecturer in Theatre at the Victorian College of the Arts and has published widely in theatre and interdisciplinary arts. Email: sarah.woodland@unimelb.edu.au

Frank Wu is an ARC Laureate postgraduate research student at The University of New South Wales's iCinema Centre. His current research primarily focuses on applying machine-learning techniques to wildfire simulation. In addition to his work in wildfire simulation, he also has interests in computer vision for recreating scenarios and reinforcement learning. Email: frank.wu1@unsw.edu.au

Chapter 1
Introduction

Dennis Del Favero, Susanne Thurow, Michael J. Ostwald, and Ursula Frohne

As global warming accelerates, climate events such as extreme fires and floods are escalating in their unpredictability, intensity and frequency, taking lives, destroying communities and wreaking havoc on habitats. The scope, scale and uncertainty of these extreme events, and their impacts on human populations, are demolishing commonly held beliefs about safety, cultural norms and emergency protocols. Preparing for these threats at the community level, not simply mitigating and managing their impact, is an increasingly existential task. Without the ability to experientially preview and address these unforeseen vulnerabilities in situ, it is extremely challenging for researchers, frontline personnel and local communities to understand their likelihood, let alone engage with practices that minimise the devastation they entail to build the imaginative readiness required.

In this context, the ability of visualisation to communicate vulnerabilities quickly and powerfully has long been recognised and is being drawn upon in heterogeneous fields. The defence and mining sectors, for example, are using immersive visualisation for scenario training as it enables personnel to feel as if they are present in virtual scenarios where they can physically experience dangers and safely rehearse their response to threats in advance. Aesthetic experimentation in the arts and sciences is increasingly being applied to visualise the complex experiences and data

D. Del Favero (✉) · S. Thurow
iCinema Centre for Interactive Cinema Research, The University of New South Wales, Kensington, Australia
e-mail: d.delfavero@unsw.edu.au; s.thurow@unsw.edu.au

M. J. Ostwald
School of Built Environment, The University of New South Wales, Kensington, NSW, Australia
e-mail: m.ostwald@unsw.edu.au

U. Frohne
Centre for Advanced Study, Universität Münster, Münster, Nordrhein-Westfalen, Germany
e-mail: ursula.frohne@uni-muenster.de

D. Del Favero et al. (eds.), *Climate Disaster Preparedness*, Arts, Research, Innovation and Society, https://doi.org/10.1007/978-3-031-56114-6_1

embodied in climate scenarios. However, these are constrained by disciplinary conceptualisations and practices that centre on data modelling or evocative impact. In climatology, for example, extreme events are visually formulated as an atmospheric disturbance, in engineering as infrastructural risk, in ecology as environmental damage, in the social sciences as adaptive vulnerability and in the creative arts and humanities as affective loss. These result in detached depictions that tend to render humans as passive observers of the aftermath and the events as the effects solely of natural laws. They are unable to viscerally embody the uncertain interactions between populations and constantly fluctuating location-specific climate variables such as intense ember storms and torrential rains.

As the totemic 1.5 °C climate target is breached and a predicted 2° if not 3° horizon approaches, this requires a complete reframing of visualisation. It demands leveraging the arts and technology to shift the formulation from impersonal depiction and detached prediction to an embedded visualisation that enables stakeholders to prepare by viscerally engaging with unforeseen geolocated threat scenarios in advance. Such integration can deliver the experiential plausibility facilitated by the arts and the physical probability enabled by technology and science. It would transform the current singularity of disciplinary approaches by integrating lived experiences and abstract data into compelling dramatisations that enable stakeholders to sensorially feel they are present in local scenarios where they can interactively experience and respond to probable vulnerabilities. Such rehearsal of climate scenarios could enhance imaginative threat perception, visceral decision making, situational awareness and adaptive readiness using virtual scenarios. Virtual scenarios have been proven to be powerful ways of envisioning complex interactions and pathways. They are experientially multidimensional and open-ended, allowing the multiplicity of flexible interpretations and responses required in such dynamic situations.

The foundations for such scenario-driven virtual visualisations are already in place. Recent advances in digital arts are able to render previously unimagined virtual worlds in extremely high fidelity, and developments in climate science facilitate visualisation of geolocated extreme event micro-climates, while generative AI and machine learning breakthroughs are being used to analyse fire and flood data and then model and construct more accurate simulations.

Integrating these advances will empower stakeholder communities to viscerally imagine threats, enact hazardous stories, mock up risk-laden transactions and probe hyperlocal preparedness. Such advances will transform visualisation's use from *what will happen* to *what it could look like* and *rehearsing how to prepare*. It will establish a new multidisciplinary domain of 'climate scenario visualisation'. Reframing the envisioning of climate disasters from the impersonal viewing of unexpected incidents into an experiential preview of possible encounters will enable communities to prepare for unforeseen emergencies through the immersive rehearsal of compelling data-driven scenarios that bridge the fictional and factual. In an era of increasing terrestrial uncertainty, such a paradigm enables us to reimagine extreme event encounters, so we can increase and tailor preparedness.

Drawing on the multidisciplinary expertise of 38 senior and emerging leaders in the creative arts, climate sciences, environmental engineering and intelligent systems, this book canvasses advances and considers how developing knowledge and practice may be integrated and leveraged to realise an innovative multidisciplinary paradigm. Grouped into four sections, the 14 chapters in this book explore the emerging strengths and current disciplinary limitations in addressing the challenge of envisioning unpredictable extreme event interactions from theoretical and experimental perspectives.

The first part, 'Picturing', charts developments in digital art, data modelling and intelligent simulation that could be utilised to envision situations that explore imaginative risk perception.

The second part, 'Narrating', surveys current performative arts that could be deployed to enhance previewing narratives that embody visceral decision making.

The third part, 'Rehearsing', explores design settings that could be adapted to mock up interactions that test situational awareness.

The fourth part, 'Communicating', canvasses lessons from recent inquiries to conceptualise how geo-specific scenarios could model experiences that probe readiness.

The contributions in this book provide a timely intervention into the global discourse on how art, culture and technology can address the climate emergency. It proposes a pioneering approach that meshes advances in the arts with those delivered through technology and the sciences to address the overlooked challenge of *climate disaster preparedness*. As such, the book appeals to readers from multiple fields, offering academic, industry and lay readers novel insights into a dangerously under-researched area in the current knowledge landscape. Addressing these challenges demands a paradigm shift from impersonal observational depictions to an AI-generated creative framework that enables stakeholders to sensorially animate and interact with geolocated physics-based scenarios, both historical and probable. This would empower communities to imaginatively and practically rehearse these life-threatening encounters in advance and transform their readiness in the face of the unforeseen.

Part I
Picturing

Chapter 2
Reimagining Extreme Event Scenarios: The Aesthetic Visualisation of Climate Uncertainty to Enhance Preparedness

Dennis Del Favero, Susanne Thurow, Maurice Pagnucco, and Ursula Frohne

Abstract Responding to the rapidly escalating climate emergency, this chapter outlines transformative multidisciplinary research centred on the visualisation of unpredictable extreme event scenarios. It proposes a unique, scientifically grounded artistic approach to one of the world's most immediate challenges—preparing communities for extreme climate events, such as firestorms and flash floods. As preparedness is a function of prior threat experience, it argues that visualising threat scenarios in advance can be a key to surviving and adapting in an era of increasing climate instability. This approach can enable communities to viscerally experience and rehearse threat perception, situational awareness, adaptive decision making and dynamic response to unexpected life-threatening extreme events. Using experimental case studies at The University of New South Wales's iCinema Research Centre and international benchmarks, this chapter explores how advances in immersive visualisation and artificial intelligence aesthetics can be integrated to provide a framework that enables the virtual prototyping of unforeseen geolocated climate scenarios to facilitate readiness in the face of accelerating climate uncertainty.

D. Del Favero (✉) · S. Thurow
iCinema Centre for Interactive Cinema Research, The University of New South Wales, Kensington, Australia
e-mail: d.delfavero@unsw.edu.au; s.thurow@unsw.edu.au

M. Pagnucco
School of Computer Science & Engineering, The University of New South Wales, Kensington, NSW, Australia
e-mail: morri@unsw.edu.au

U. Frohne
Centre for Advanced Study, Universität Münster, Münster, Nordrhein-Westfalen, Germany
e-mail: ursula.frohne@uni-muenster.de

D. Del Favero et al. (eds.), *Climate Disaster Preparedness*, Arts, Research, Innovation and Society, https://doi.org/10.1007/978-3-031-56114-6_2

Keywords Aesthetics · Artificial intelligence · Climate change · Creative arts · Immersive visualisation · Preparedness · Scenario modelling · Uncertainty simulation

2.1 Introduction

The climate emergency presents an existential global crisis resulting from the combined processes of global warming and atmospheric, hydrospheric, biospheric and pedospheric degradation. The IPCC report of 2023 found that extreme climate events are rapidly increasing around the globe, with projections indicating they will become more frequent and severe, with impacts intensifying and interacting. The World Economic Forum's *Global Risks Report 2023* identifies, for the first time, "failure to mitigate climate change" and "failure of climate change adaptation" as the most severe risks on a global scale, followed by "natural disasters and extreme weather events" (Heading & Zahidi, 2023).

In Australia, for example, natural disasters, primarily wildfires and floods, have cost *over $13 billion* every year (1.2% of GDP) and are "expected to rise to $39 billion per year by 2050" (Slatyer et al., 2017). Over nine million Australians have been directly impacted by extreme weather events since 1990. The Australian government's *National Strategy for Disaster Resilience* and its *National Climate Resilience and Adaptation Strategy* emphasise the critical importance of employing all possible means to mitigate the impacts of climate emergencies on communities.

In reference to the findings of the Royal Commission into National Natural Disaster Arrangements (Binskin, 2020), Emergency Leaders for Climate Action (2022) urged that "unprecedented is not a reason to be unprepared. We need to be prepared for the future". As detailed by Cunningham et al. in Chap. 13 of this volume, Australia's policy settings have thus far failed to facilitate a level of effective preparedness, adaptation and mitigation that can pave a way to sustaining quality of life (or life itself) on our continent. The Australian Productivity Commission (2014) reports that 97% of Australia's investment in climate emergencies is on recovery and only 3% on preparedness and mitigation. As fire scientist David Bowman (2023) points out, Australia is currently "sleepwalking" into a future of extreme and escalating climate events that will inevitably change how we can live on this "most fire-prone continent[-] on Earth". As Cunningham et al. argue, investing in preparedness is key to meeting this challenge. However, despite its importance, tangible examples of enhancing preparedness are rare, partly because it often requires deep levels of experiential and intellectual engagement across disciplinary boundaries (Lazo et al., 2015).

The climate crisis requires fundamental shifts in how these events are envisioned. Creative arts and technology are increasingly being used to depict and model these world-defining events, using evocative imagery and abstract graphics to communicate with audiences (e.g. Altintas et al., 2015, Calkin et al., 2021, Smith, 2015, Roelstraete et al., 2023). By depicting complex climate information in dynamic visual form, art and technology enable a more immediate and intuitive understanding of extreme events in ways that are not apparent in raw data (Shepherd & Truong,

2023). The arts can achieve this by depicting plausible scenarios, while technology undertakes it by modelling probable scenarios. Yet both currently tend to provide episodes that address non-localised uncertainties and render citizens as passive witnesses to the aftermaths, or the events as abstract forces, constraining the capacity to anticipate reliable scenarios (Jasanoff, 2010; Sheppard 2019). This leads to detached generic observations, whereas credible and meaningful preparedness emerges from embedded sensorial experiences that can inform decision making, since preparedness is a function of the prior experience of a perceived threat (Lazo et al., 2015). Most importantly, current art and technology approaches are unable to experientially vivify the increasingly uncertain interaction between geolocated events and situated communities, amplifying a profound existential vulnerability (Sheppard, 2005).

To overcome this challenge, government agencies, emergency services and researchers are seeking multidisciplinary innovations to optimise foresight, readiness and responsiveness to face the catastrophic risks of global warming (Jasanoff, 2010). This challenge demands reformulating how these interactions are envisioned. It necessitates a shift from abstract observational depiction and modelling to a scenario visualisation that facilitates a viscerally immersive interaction with unforeseen situated threats in advance. This formulation would need to integrate the recent advances in the arts that create compelling visceral experiences with developments in artificial intelligence (AI) that process and render complex and unpredictable climate processes.

This chapter surveys the conceptual and practical requirements for integrating these approaches and establishing a climate scenario visualisation framework to address this existential challenge. It probes preliminary experimental case studies developed at The University of New South Wales's iCinema Research Centre in tandem with international benchmark research. It outlines the state of the art in current scenario visualisation and maps how recent advances can be leveraged to address significant constraints currently limiting progress in the development of preparedness, both in understanding and practice. The chapter argues that what is required is an immersive visualisation and AI aesthetic that facilitates actionable visualisation (Sheppard, 2005) using virtual landscape scenarios in which viewers can safely curate the variables that drive extreme events and compose episodes through which to rehearse and test responses. In such dynamic landscapes, communities can viscerally and interactively enact scenarios that embody the unforeseen interactions between themselves and constantly fluctuating climate variables. Such a multidimensional aesthetic framework would act as an intelligent simulatory theatre where communities could creatively envision their transaction with the intense uncertainty and non-linearity of extreme events in their locality.

2.2 Immersive Visualisation

Recent developments in immersive visualisation enable users to navigate through dynamically modelled territories, where they can experience multisensorial vulnerabilities and safely rehearse their response to threats in advance within a credible spatial environment (e.g. Soga et al., 2021). Immersion here refers to physically embedding the user in a virtual scenario where they can navigate through its three-dimensional (3D) space, enabling them to spatially explore scenarios as if they are present (Fonnet & Prié, 2021; Suh & Prophet, 2018).

This use of immersive scenarios to simulate threat preparedness is evidenced in the *iCASTS* project (Shaw & Del Favero, 2010–ongoing) that to date has trained over 30,000 underground miners in New South Wales. It uses 360° 3D theatres situated on mine sites in which plausible naturalistic simulations of probable evidence-based hazardous settings are presented. Integrating artistic and technological methods, trainees experientially engage with predictable threats, underpinning a 65% reduction in serious injuries and no fatalities across the NSW mining sector (Pedram et al., 2014; Fig. 2.1).

In similar ways, the University of California's Center for Information Technology Research in the Interest of Society (CITRIS) is developing immersive interactive projects that focus on visualising navigational decision making during a disaster. It is their aim to enhance communications between emergency services and the public, to understand how future evacuations may be improved. Stakeholders use head-mounted VR to rehearse the navigation of expected hazards during an evacuation simulation "to help shift [their] perspectives from reacting to events when they occur, to anticipating emergencies and looking for ways to reduce risk before disaster strikes" (Soga et al., 2021).

Virtual scenarios are proving to be powerful tools for envisioning complex interactions and for probing viable pathways for action (Havenith et al., 2019). They are multimodal, combining sight, sound and kinaesthetic dimensions, and

Fig. 2.1 Installation view of *iCASTS*, iCinema (2023)

multidimensional, enabling a multiplicity of flexible responses that are required in chaotic situations (Lempert, 2013). However, current approaches lack the ability to depict unpredictable interactions between stakeholders and geolocated events, where uncertainty is central. This severely limits their capacity to support preparedness.

Addressing these interactions demands developing immersive scenarios that model rapid, large-scale and unanticipated transactions in geolocations, which cannot be understood by human cognition alone. It requires a transformative aesthetic that integrates cutting-edge advances in creative arts and AI. This would combine the speed and scale of AI in establishing patterns and predicting behaviours, with the digitally augmented inventiveness of aesthetic practices to make sense of and process the uncertainty of situated sensorial experience (Grosz, 2001; Del Favero et al., 2023). The complex nature of climate scenarios requires a multilayered aesthetic framework that utilises different forms of sense making—encompassing picturing a situation, accounting for what is seen, communicating what is experienced and rehearsing a response (Lempert, 2013).

Such a framework would empower stakeholder communities to viscerally perceive threats, dramatise hazardous stories, prototype risk-laden encounters and probe geolocal readiness. In short, it would mobilise preparedness by previewing and rehearsing an appropriate response. Reframing visualisation from *what will happen* to *what it could look like and how to prepare*, it will transform the art and technology involved and establish a transformative knowledge domain in climate scenario visualisation. This will shift the envisioning of climate disasters from the passive observation of events into an experiential preview of hyperlocal unexplored situations (Frohne, 2023). It will enable readiness and adaptive responsiveness for unforeseen emergencies through the embedded rehearsal of probable and plausible geolocated scenarios. Investing in such a risk reduction strategy will provide a compounding dividend of avoided loss and suffering, reduced disaster costs and enhanced creative capacity, cultural cohesion and social empowerment.

Such an integration of art and technology needs to start by reconsidering how we imaginatively formulate these crises and our interaction within them. As set out by Thurow, Grehan and Pagnucco in Chap. 9, dynamic systems theory, such as that of Bruno Latour (2018), considers humans as part of a symbiotic "terrestrial habitat". The concept of the "terrestrial" redefines humans as one of the many Earth-bound agents, co-habiting with multiple other organic and non-organic agencies, including the sensorial and cognitive aesthetic processes and forces that form part of the habitat. The climate breakdown is disrupting these terrestrial and corresponding aesthetic systems. This instability is now centre-stage and manifested in the uncertainty across all terrestrial systems and the ways they are aesthetically formulated (Willcock et al., 2023). It triggers intense climate variability, multiplicative stresses and intersystem feedback with indeterminate outcomes—as seen, for example, in the unexpectedly accelerated rise of phosphorus concentrations in freshwater reserves (ibid.). This increasing uncertainty is challenging predictive data modelling and existing imaging paradigms, necessitating an aesthetic that can address the resulting systemic instability. As Britts, Song and Cortes discuss in Chap. 6, climate scientists are proposing aesthetic modifications to simulation approaches that focus on situated scenario visualisation to augment decision making, using storylining (Shepherd et al., 2018), tales (Hazeleger et al., 2015) and scenarios (Lempert, 2013).

Such a situated scenario visualisation would need to be grounded in robust data to ensure it aligns as accurately as possible with the determinate and indeterminate physical processes that govern extreme events. As Moinuddin et al. and Song et al. demonstrate in Chaps. 4 and 5, respectively, new physics-based modelling supported by machine learning (ML) and generative AI can supply such a basis if coupled with efficient processing pipelines. However, by themselves, data analysis and established visualisation approaches are currently not capable of furnishing meaningful insight and engagement with situated extreme event modelling. They yield vast amounts of data that are most commonly visualised through abstract graphical illustrations—struggling and often failing to convey geolocated dimensions of the events, which in their immediacy directly influence people's response in emergencies. The visceral experience of a fire, its terrifying unpredictability and sensorial scale, can easily override any preparatory plans and actions devised via abstract graphics or text-based engagement.[1] To ensure people have a safe and reliable concept of what to expect when facing a firestorm or flash flood, they need to be presented with a compelling and credible rendition of what facing such events might look and feel like. Algorithmic visualisation systems can be complemented by creative methodologies so that people may experience such overwhelming scenarios in advance in a safe virtual environment, allowing the chaos of the situation to be moderated through visceral rehearsal of response strategies. As Grehan, Ostwald and Smaill argue in Chap. 14, the arts and architecture excel at aesthetically transmitting such experiential qualities and can do far more than merely facilitate evocative stories of climate change impact. They can provide the key to unlocking an affective and intelligent preparedness based on enhanced data insight, visual analytics and rehearsal approaches. These can transform isolated disciplinary approaches into a cohesive and powerful integrated approach.

2.2.1 AI Aesthetics

Actively involving viewers in the composition of climate scenarios can significantly enhance their immediacy and meaning (Stevens et al., 2023). Traditional linear narrative concepts derived from semiotic and semantic theories are ill-suited to realise such a dynamic formulation as they fail to leverage the expanded capabilities of immersive visualisation and AI to reformulate engagement with the pillars of storytelling, namely, progression through time and space (Deleuze, 1995). Narratives that adhere to realist paradigms that simulate everyday experiences of these dimensions are limited in their capacity to stimulate reflection and to afford insight into

[1] That is, those from emergency responders and local communities

the fundamental relations that embed us in our habitat. While hyperrealistic renditions of pre-scripted scenarios, such as those provided in Belinda Chayco and Tony Ayres's seminal TV series *Fires* (2021), can impress upon audiences the ferocity of violent firestorms, they cannot facilitate personally relevant foresight into viable response strategies when faced with such incidents. This is because the linear progression that is determined by the traditional televisual medium already prefigures the causal logic and semantic valuation of selected narratives and frames a passive viewing position. By contrast, the malleability that underpins immersive visualisation and AI aesthetics allows transforming such detached positioning by affording the means to radically redefine narrative forms and interactions.

Digital artefacts are multimodal in nature, "shaped by software rather than semiotic codes: [that is,] software compresses information into virtually realisable [and interpretatively] thick units" (Weibel, 2002). Rather than applying psycholinguistic approaches that understand narrative as the recovery of representational structures from semantic memory (Willemen, 2002), digital media are better served by deploying a concept of narrative defined "as the episodic recomposition of emergent events within the affective, sensory and cultural memory" of the viewer (Deleuze, 1995). Such a definition captures and engages the layered and autonomous status of data in its virtual form, which is open to actualisation in manifold combinations and translation across varied contexts (Manovich, 2001). It develops concepts of relational semantics that have conceptually transferred meaning-making authority from author to receiver of information, emphasising that the aesthetics of data carry their own implications for aesthetic practice that are activated in dialogical engagement with the viewer. Narrative in these contexts becomes recombinatory, defined as a recursive system made up of a large number of self-organising and interdependent data elements that are able to provide the viewer with richly textured engagements. As a technical and critical framework, a recombinatory narrative recycles the abundance of available information into significant multitemporal episodes. For example, rather than presenting viewers with a predetermined set of narrative permutations, they can actively partake in combining and curating units of data into open-ended and infinitely malleable storylines. Applied to flood visualisation, this insight into and engagement with the dynamics of these events is far better enabled by letting viewers actively explore the effects of flood variables and evacuation decisions on the progression of a scenario than if they merely watched a recording of the events as they unfold.

The effectiveness of such approaches has already been foreshadowed by international seminal research, such as Refik Anadol's *Archive Dreaming* (2017), which uses an AI framework to enable a user to search and sort 1.7 million archival documents and convert these into dynamic and novel narratives within an immersive theatre. When idle, the installation generates unexpected correlations between the documents. In a similar vein, the *T_Visionarium* project (Del Favero et al., 2004–2009) offers the means to capture and reassemble televisual data supported by an AI image analysis and recombinatory system within a panoramic 360° theatre. It allows viewers to explore and actively compose a multitude of stories reassembled from the original data. Digital free-to-air Australian television was captured over a

1-week period. This footage was segmented and converted into a database containing over 20,000 short clips. Each clip was first manually tagged with metadata descriptors defining their properties and then processed using AI. The AI image analysis and recombinatory system assembles and displays across the 360° screen a selection of related visual material based on the metadata. The latter includes categories such as the gender of the actors, the dominant emotions they are expressing and the prevalent colouring of the scene. Dismantling the video data in this way breaks down the original linear narrative into components that then become building blocks for a new kind of interactive television. Two hundred fifty video clips are simultaneously displayed and distributed around the panoramic theatre. The user can select, rearrange and link these video clips at will, composing them into combinations based on relations of gesture and movement. They are able to reassign connections among data layers by pleating and creasing their topology until they cascade into new episodes of autonomously unfolding events. The AI furnishes the user with multiple entry and exit points to the data, with the facility to generate narrative content on the fly (Manovich, 2001). Experiments such as these do not attempt to freeze the world into monolithic representations. Instead, they explore how the world may be dramatised in ways that shed light on how it is an assemblage of images and processes (Serres, 2000; Fig. 2.2).

T_Visionarium explores the productive capabilities of recombinatory narrative: that is, deconstructing televisual sequences into their building blocks and allowing their interactive recomposition, to probe new creative modes of user-driven storytelling. When seeking to make an interactive recombinatory system useful for

Fig. 2.2 Exhibition view of *T_Visionarium*, iCinema (2023)

exploring climate processes, fully open-ended user control misaligns with the task of plausibly engaging with physical laws. User domination could easily override these and fail to account for the global-scale interactions that shape extreme events. These systems stand in continuous exchange with each other as well as with human and non-human agents (Latour, 2018). To develop compelling scenarios while retaining plausibility requires integrating artistic free-form with a grounding in real-world physics. Users have to feel empowered to affect the visual environment while at the same time encountering the resistance that conveys the actions of non-human agencies. If such resistance is perceived as coherent, i.e. following discernible patterns, then a dialogue between users and an interactive system can be established and explored.

The *Atmoscape* project (Del Favero et al., 2012–2014) embarked on such an experimental investigation by processing low earth-orbiting satellite data of lower atmosphere water vapour provided by NASA. This was translated into immersive scenarios for scientific and artistic applications. For the scientific application, the immersive visualisation allowed scientists to observe and interactively study phenomena such as reticulations in water vapour layers preceding tropical cyclone formation for the very first time, as they were undetected in conventional raw data modelling. In its artistic application, the *Nebula* project (2014–2023) investigated the emergence of an aesthetic that demonstrated the potential for alternative conceptualisations to the pervasive idea of landscape, including its weather and atmosphere, as an inert backdrop to human activity. Driven by an AI particle-generation graphics engine, *Nebula* allows users to interact with water vapour particles underpinning a virtual landscape. They can interactively assemble resistant AI-programmed particles into a range of clustering topographies and vistas. While immersed in these vertiginous terrains, users hear a voice challenging them to explore unknown sites. As they attempt to herd the independently minded particles into recognisable landscapes, the undulating generative imaging system suggests how their actions and those of the landscape are fundamentally enmeshed, co-dependent yet autonomous. With particles constantly shifting and re-assembling, giving way to new combinations, users are forced to engage with these dynamic processes and to find ways of navigating and responding to the independent kinaesthetic flux of particle streams. In its abstraction of atmospheric data and its encoded processes, *Nebula* affords users the opportunity to explore an AI aesthetic that is underpinned by analysis and embodiment of water vapour dynamics. Such an AI aesthetic can drive insight into the constitution of terrestrial systems as dynamically mutating situated habitats (Fig. 2.3).

In order to furnish an evolving transaction in which both human and other agencies mutually determine each other, capabilities for such dialogical interaction need to be developed that articulate how agencies are symbiotically asymmetrical. Narrative has to co-evolve between human and virtual agents based on autonomous yet reciprocal perception and interpretation of behaviours and interactions. The *Scenario* project (Brown et al. 2011–2015; Scheer, 2012) experimentally investigated such forms of co-evolutionary relationship between virtual agencies and human participants in immersive environments.

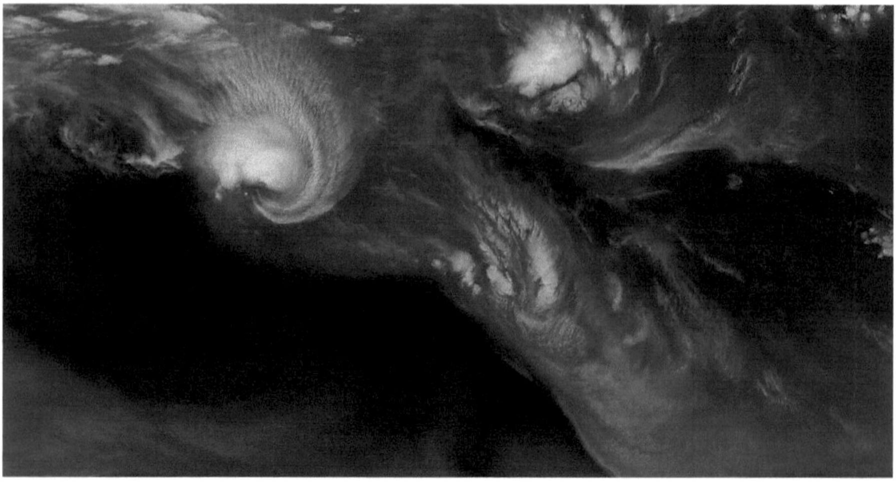

Fig. 2.3 Exhibition view of *Atmoscape*, iCinema (2023)

In *Scenario*, a female humanoid character and her children seek to flee an underground labyrinth, enlisting the users' help to identify viable escape routes. A group of shadowy AI-enabled humanoid sentinels are tracking the family and users, trying to block their attempts at successfully navigating the nested spaces. The sentinel interaction is achieved by means of a computer vision system that tracks the users' behaviour, linked to an AI system that allows the humanoid virtual agents to independently interpret and respond to user behaviour—with actions sampled from a knowledge database. The AI system was developed using a variant of a symbolic logic planner drawn from the cognitive robotics language Golog, capable of dealing with sensors and external actions. Animations that can be performed by a humanoid character were considered actions that needed to be modelled and controlled.[2] Each action was modelled in terms of the conditions under which it could be performed[3] and how it affected the environment when the action was performed. Using such modelling, the AI system planned and coordinated the actions (i.e. animations) of the humanoid characters by reasoning about the most appropriate course of action. This imbued the sentinels with a number of capacities beyond the rudimentary prescripted symmetrical behaviour of conventional virtual agents such as regular computer game characters. First, they were invested with the ability to sense the behaviour of individuals as well as collectives of users. Second, the AI system enabled symbolic representation of this behaviour. And third, the agents were able to deliberate on their own behaviour and respond intelligibly through gestural and clustering actions. The framework was structured so as to respect autonomous virtual agent intentionality, as opposed to the simulated intentionality of conventional

[2] For example, walking to a location, pushing a character, etc.

[3] For example, pushing something or someone could only be executed if located next to an object or agent.

Fig. 2.4 Exhibition view of *Scenario*, iCinema (2023)

games. While narrative reasoning in human-centred interactivity focuses exclusively on human judgements, the co-evolutionary narrative here allows deliberate autonomous action by virtual agents (Scheer, 2012). *Scenario* was designed to dramatise distinct behavioural processes, thus probing virtual agent autonomy and the cognitive gap between virtual agents and human participants. It investigates the differences in narrative reasoning between them. It probes how virtual agents, if provided with a modest ability to sense and interpret the actions of human users in a shared immersive environment, can interactively respond and co-evolve autonomously with the users (Fig. 2.4).

These AI-enabled capabilities provide a robust foundation for the exploration of evolving and cascading interaction between autonomous climate agencies and human participants in an immersive environment. This is the prerequisite for conceiving open-ended and unimagined encounters that plausibly anticipate probable extreme events.

2.3 Towards Climate Scenario Visualisation

iFire (Del Favero et al., since 2021) explores the visualisation of dynamic wildfire scenarios. It is an experimental prototype for climate scenario visualisation, integrating the advances in immersive visualisation and AI aesthetics in the above-cited case studies. It is being developed in collaboration with a range of national and international partners, including the Australian Broadcasting Corporation, AFAC (Australasian Fire and Emergency Service Authorities Council), CSIRO/Data 61, Fire and Rescue NSW, Royal College of Art London, San José State University's Wildfire Interdisciplinary Research Center and The University of Melbourne. It assembles a team of Australian, European and US artists; AI, computer, fire and climate scientists at international universities; and partner organisations to develop immersive visualisations of extreme wildfires and their uncertain dynamics. It utilises an AI-based landscape prototype that not only interprets but learns from human interaction and behaves autonomously. It sketches a scenario visualisation system that can depict unpredictable climate event interactions.

iFire's goals are to:

1. Establish synthetic landscapes that can envision unpredictable wildfires and explore imaginative risk perception.
2. Create improvised narratives that can dramatise unanticipated fire behaviour and embody visceral decision making.
3. Enact interactive experiences in rural settings where users can virtually rehearse unexpected fiery encounters.
4. Generate geolocated scenarios that can model unanticipated fire-laden landscapes and probe readiness.

The *iFire* prototype is being developed for application as an artistic and a scientific series, respectively titled *Penumbra* and *Umbra*. It consists of a database of atmospheres, flora, pyro-histories and topographies and AI landscapes. To ensure geophysical reliability, it uses both the SPARK and WRF-SFIRE simulation engines applied to geographical databases. To implement a rich sensorial credibility for these worlds, it uses the Unreal game simulation engine and a customised interface specific to each series. This enables higher-fidelity texturing and modelling of the landscape at scale. The interfaces are venue domain-specific, such as for museums, scientific laboratories and emergency centres. For artistic applications, it utilises cinematic environments where interaction is driven by motion tracking. For science and emergency applications, it uses either a tablet for physical screen environments or mouse for online environments. A main window display is used for landscape navigation, with smaller inset informational windows to convey variables such as wind speed. Each series is being developed in collaboration with domain-specific stakeholders. Both series are explored through three geolocated case studies: a hypothetical pine plantation fire, a grasslands fire in the Australian state of Victoria (2022) and the Bridger Foothills Fire in Montana, USA in 2021. The case studies progressively explore the intensifying dynamics of fire from the irregular rhythms of a low-intensity fire to the violent vorticity-driven lateral spread of an extreme fire.

The artistic *Penumbra* series explores the multisensorial qualities of unexpected wildfire experiences for creative industry audiences. The scientific *Umbra* series investigates the dynamic interplay between unanticipated wildfire processes and users for scientific analysis and emergency training. Both series are remotely accessible. They are developed across a networked, fully immersive system that can translate multilayered wildfire data into 3D scenarios, allowing exploration of situations across distributed locations. This comprises immersive cinemas, wall displays, desktops and tablets. It generates hyperrealistic immersive visualisations that can stage a geolocated wildfire as it unfolds on the ground in interaction with users. The visualisations operate in two modalities—the first enables users to create compelling hypothetical scenarios, and the second allows users to recreate historical fires and generate probable future scenarios.

The AI system that underpins the wildfire landscape has three distinct functions. First, its ML system addresses shortcomings in existing empirical and physical models. For physical models, the computation complexity grows exponentially with higher resolution and takes an inordinate amount of time to simulate, which is prohibitive for operational use. On the other hand, empirical models are mostly used in operations because of their simplicity. However, due to simplified modelling of the processes and dynamics of wildfires, empirical models often fail to achieve the accuracy required. To address the disadvantages of both physical and empirical simulations, *iFire*'s ML-based simulations learn multidimensional representations of data in latent space and apply physics-related constraints to a neural network. Once trained, it provides instant and accurate inference.

Second, the AI system implements algorithms for the interpretation of motion capture and tracking data derived from individual users and groups. It applies algorithms to enable the AI landscape to set goals, learn from and autonomously respond to user behaviour. This furnishes the landscape with the ability to develop and realise its goals while integrating what it learns from human behaviour through exceptionally fast execution enabled by the programming language verified in the *Scenario* and *Nebula* projects. This interprets past user behavioural patterns, models and anticipates future behaviour and makes inferences about how to act in response. It is achieved by using open-ended rules that allow the AI setting to infer from unknown situations, make predictions, decide how to act, learn from the interaction and adapt its reasoning. This ensures that the setting can make independent decisions, a critical attribute for its improvised and reciprocal interaction with users so that it acts in unanticipated ways. Third, the AI goal-orientated system analyses user interactive decision making to both support and challenge it, in order to optimise user response to scenario uncertainty.

The AI is applied in different ways for the two series. The artistic *Penumbra* series leverages AI to explore an open-ended user and landscape dialogue that co-evolves through reciprocal transactions between the user and the fire-laden terrain. For example, the landscape may choose to collaborate with the user by enlivening the landscape and propagating fire-resistant trees, or it may challenge the user's attempt to control its spread by generating chaotic ember storms. The unpredictable nature of its behaviour opens pathways for new forms of interactive encounter that are open to user and AI co-creativity while circumscribed by physics. By translating wildfire behaviour into discrete actions, *Penumbra* facilitates an experimental imaginary where possibilities and vulnerabilities can be generated and investigated. Opening a new horizon for reimagining extreme fires, it enables prototyping an anticipatory life-saving imaginary. By interactively exploring and transforming the

Fig. 2.5 Exhibition view of *Penumbra*, iCinema (2023)

dynamic range of possibilities involved, it offers artists and audiences a new genre of collaborative human and machine co-creativity through which they will be able to compose a wide spectrum of previously unforeseen and mutating encounters with evolving fire landscapes unrestricted in aesthetic form, complexity and inventiveness (Fig. 2.5).

The scientific *Umbra* series leverages the AI to furnish an analytical laboratory for scientists and a training platform for emergency personnel, such as fire crews, incident controllers and operations officers (Fig. 2.6). The focus for the first is to facilitate 3D animated models of extreme fires in geolocated landscapes where scientists can either forensically assess historical fires or test their hypotheses for upstream scenarios by manipulating variables such as wind speed and fuel load. For the second, the focus is to train emergency service personnel's situational awareness and decision making in a safe environment. Actual wildfire grounds are impossible to use as training environments due to their clear and present dangers. Emergency personnel often have very short timespans to spot key indicators of impending conflagrations and to make life-saving decisions. This series allows personnel to adjust the situational variables to focus on a specific critical variable, such as wind direction, and to viscerally experience the dramatic changes these can trigger. They are able to explore and practise the efficacy of a range of different responses, training their perceptual skills. This allows them to momentarily step away from the hyper-attention required in the real-life situation, to rethink, reflect and reconsider the complex relationships that structure and govern their action space. The AI system

Fig. 2.6 Screen view of *Umbra*, iCinema (2023)

learns from, predicts and disrupts their reactions while presenting a challenging range of novel fire behaviours. These features allow emergency personnel to experientially deal with unpredictable scenarios and develop proactive planning through dramatisations that safely simulate complex uncertainties. Groups of responders are physically placed inside a rapidly moving wildfire landscapes and confronted with evolving situations that challenge them to collaborate across geographic locations. Emergency organisations can integrate location-specific data and protocols into the simulation to build informational complexity and provide trainees with the challenging and unanticipated experiences they need to manage and mitigate risk.

2.4 Conclusion

As extreme event preparedness is a function of prior threat experience, safely visualising threat scenarios in advance is key to enhancing survival and adaptation in an era of unpredictable extreme event emergencies. A climate scenario visualisation framework aims to model a visceral imaginary to foster community preparedness by aesthetically integrating and transforming artistic, technological and scientific approaches. It would generate the capacity to make sense of an extreme event by picturing the situation, narrating its contours, interacting with its dynamics, communicating its experiences and testing a credible response. This would enhance a community's ability to viscerally experience and develop threat perception, situational awareness, adaptive decision making and flexible response to unexpected life-threatening situations. By integrating advances in immersive visualisation and AI aesthetics, such a framework would virtually rehearse unforeseen geolocated extreme events to facilitate readiness in the face of escalating and profound climate uncertainties.

Acknowledgements The research described in this chapter has been supported by the Australian Government through the Australian Research Council's Laureate and Discovery funding schemes (FL200100004, *iFire*; DP120102243, *Atmoscape*; DP110101146, *Nebula*; DP0556659, *Scenario*; DP0345547, *T_Visionarium*). All authors are ARC investigators as listed for funding scheme). This chapter includes extracts by the authors from the iCinema website.

References

Altintas, I., Block, J., De Callafon, R., Crawl, D., et al. (2015). Towards an integrated cyberinfrastructure for scalable data-driven monitoring, dynamic prediction and resilience of wildfires. *Procedia Computer Science, 51*(1), 1633–1642.

Anadol, R. (2017). *Archive dreaming*. Exhibited at: San Francisco, USA; Istanbul, Turkey. https://shorturl.at/biBL8. Accessed 13 Dec 2023.

Ayres, T., Chayko, B., Denholm, A., & Watts, L. (executive producers). (2021). *Fires* [TV series]. Australian Broadcasting Corporation.

Binskin, A. C. M. M. (2020). *Royal Commission into national natural disaster arrangements report*. Commonwealth of Australia.

Bowman, D. (2023). 'Australia is sleepwalking': A bushfire scientist explains what the Hawaii tragedy means for our flammable continent. *The Conversation*. https://shorturl.at/chLSX. Accessed 1 Dec 2023

Brown, N. C., Del Favero, D., Shaw, J., & Weibel, P. (2004–2017). *T_Visionarium*. Exhibitions have included: 2017: 'Art of immersion', ZKM Media Museum, Karlsruhe, Germany. 2014: Chronus Art Center, Shanghai, China. 2010: 'STRP Festival', Strijp R Klokgebouw, Eindhoven, Netherlands. 2009: 'International Architecture Biennale', Zuiderkirk, Amsterdam, Netherlands; 'Un Volcan Numérique', Le Havre, France; 'Second Nature', Aix-en-Provence, France. 2008: 'eARTS Festival: eLANDSCAPES', Shanghai Zendai Museum of Modern Art, China; 'Biennial of Seville', Spain; 'Sydney Festival'. 2007: 'YOUser', ZKM. 2006: 'Artescienza: La Rivoluzione Algoritmica, Spazio Deformato', Casa dell'Architettura, Rome, Italy. 2005: 'Avignon Festival', France. 2004: 'Cinémas du Futur, Lille 2004 Capitale Européenne de la Culture', Centre Euralille, Lille, France.

Calkin, D. E., O'Connor, C. D., Thompson, M. P., & Stratton, R. (2021). Strategic wildfire response decision support and the risk management assistance program. *Forests, 12*, 1407.

Del Favero, D., with Davidson, J. W., Green, C., Moinuddin, K., Ostwald, M. J., Pagnucco, M., & Song, Y. since 2021. *iFire*. Exhibited at: 2023: 'SIGGRAPH Asia', ICC, Sydney, 'Viscera', Cavallerriza Reale, Turin, Italy; 2022: 'Beyond the Night', 'Art Cologne' and 'Dusseldorf Cologne Open', Galerie Brigitte Schenk, Cologne.

Del Favero, D., Shaw, J., Benford, S., & Goebel, J. (2011–2015). *Scenario*. Exhibited at: 2014-15: 'Jeffrey Shaw & Hu Jieming Twofold Exhibition', Chronus Art Center, Shanghai, China; 'Child, Nation & World Cinema Symposium', UNSW, Sydney. 2013: 'ISEA', UNSW, Sydney.

2011: 'Sydney Film Festival', Sydney. 2010: '15th Biennial Film & History Conference', UNSW, Sydney.

Del Favero, D., Bennett, J., Brown, N., Shaw, J., & Weibel, P.. (2014–2023). *Nebula/Atmoscape*. Exhibitions have included: 2023: 'Viscera', Recontemporary, Turin, Italy; 2019: 'SIGGRAPH Asia', Gallery of Contemporary Art, Brisbane. 2018: 'Art Abu Dhabi', UAE; 'SIGGRAPH Asia', Tokyo International Forum, Japan; 'Visibility Matrix', Douglas Hyde Galley, Dublin & Void Gallery, Derry, Ireland; 'Art Cologne', Germany; 'Le Printemps de Septembre', La Fondation Espace Écureuil, Toulouse, France. 2017: 'Art of Immersion', ZKM Media Museum, Karlsruhe, Germany. 2016: 'Sydney Film Festival'; 'Art Cologne', Germany; Galerie Marion Scharmann & Laskowski, Cologne, Germany. 2015: 'GLOBALE', ZKM, Karlsruhe, Germany.

Del Favero, D., Thurow, S., Frohne, U., Moinuddin, K., Sharma, A., & Song, Y. (2023). Re-imagining the climate emergency using AI visualisation. In *RE:SOURCE—10th international conference on the histories of Media Art, Science & Technology*. Università Ca'Foscari Venice. September 15.

Deleuze, G. (1995). *Negotiations 1972–1990*. Columbia University Press.

Emergency Leaders for Climate Action (ELCA). (2022). *Statement: Dangerously unprepared for climate-fuelled disasters. Emergencyleadersforclimateaction.org.au*. https://shorturl.at/lu035. Accessed 9 Oct 2023.

Fonnet, A., & Prié, Y. (2021). Survey of immersive analytics. *IEEE Transactions on Visualization & Computer Graphics, 27*(3), 2101–2122.

Frohne, U. (2023). Night drifts above the Earth. In I. P. di Persano (Ed.), *Dennis Del Favero. Viscera* (pp. 28–31). Recontemporary.

Grosz, E. (2001). *Architecture from the outside*. MIT.

Havenith, H.-B., Cerfontaine, P., & Mreyen, A.-S. (2019). How virtual reality can help visualise and assess geohazards. *International Journal of Digital Earth, 12*(2), 173–189.

Hazeleger, W., van den Hurk, B. J. J. M., Min, E., van Oldenborgh, G. J., Petersen, A. C., Stainforth, D. A., Vasileiadou, E., & Smith, L. A. (2015). Tales of future weather. *Nature Climate Change, 5*(2), 107–113.

Heading, S., & Zahidi, S. (2023). *Global risks report*. World Economic Forum.

Jasanoff, S. (2010). A new climate for society. *Theory, Culture & Society, 27*(2–3, 233), –25.

Latour, B. (2018). *Down to earth*. Polity.

Lazo, J., Bostrom, A., Morss, E., Demuth, J., & Lazrus, H. (2015). Factors affecting hurricane evacuation intentions. *Risk Analysis, 35*(10), 1837–1857.

Lempert, R. (2013). Scenarios that illuminate vulnerabilities and robust responses. *Climactic Change, 117*, 627–646.

Manovich, L. (2001). Post-media aesthetics. In D. Del Favero & J. Shaw (Eds.), *disLOCATIONS*. ZKM.

Pedram, S., Perez, P., & Palmisano, S. (2014). Evaluating the influence of virtual reality-based training on workers' competencies in the mining industry. In A. G. Bruzzone, F. De Felice, M. Massei, Y. Merkuryev, A. Solis, & G. Zacharewicz (Eds.), *13th international conference on modeling and applied simulation* (pp. 60–64). Curran.

Productivity Commission. (2014). *Natural disaster funding arrangements* (Inquiry report no. 74). Canberra. JEL code: H77, H84.

Roelstraete, D., Mainetti, M., & Mattiacci, C. (2023). *Everybody talks about the weather*. Fondazione Prada.

Scheer, E. (2012). *Scenario*. ZKM.

Serres, M. (2000). *The birth of physics*. Clinamen.

Shaw, J., & Del Favero, D. (2010–ongoing). *iCASTS*. Exhibited at: Shenyang Research Institute, Fushun, 2012; APPEA Conference, Convention & Exhibition Centre, Brisbane, 2010; Mines Rescue Pty Ltd Australian Facilities in Woonona, Argenton, Lithgow & Singleton, 2008-ongoing; SimTech Conference, Convention Centre, Melbourne, 2008.

Shepherd, T., & Truong, C. H. (2023). Storylining climes. In D. Yu & J. Wouters (Eds.), *Storying multipolar climes of the Himalaya, Andes and Artic* (pp. 157–183). Routledge.

Shepherd, T., Boyd, E., Calel, R., Chapman, S. C., et al. (2018). Storylines: An alternative approach to representing uncertainty in physical aspects of climate change. *Climatic Change, 151*, 555.

Sheppard, S. (2005). Landscape visualisation and climate change: The potential for influencing perceptions and behaviour. *Environmental Science & Policy, 8*, 637–654.

Slatyer, J., Harmer, P., Callaghan, J., Ronnenberg, R., O'Sullivan, P., & Hartzer, B. (2017). *Building resilience to natural disasters in our states and territories*. Deloitte.

Smith, T. (2015). World picturing in contemporary art: The iconogeographic turn. *Australian and New Zealand Journal of Art, 7*(1), 24–46.

Soga, K., Comfort, L., Zhao, B., Lorusso, P., & Soysal, S. (2021). *Integrating traffic network analysis and communication network analysis at a regional scale to support more efficient evacuation in response to a wildfire event*. UC Office of the President: University of California Institute of Transportation Studies.

Stevens, B., Adami, S., Ali, T., Anzt, H., et al. (2023). Earth virtualization engines (EVE). *Earth System Science Data Discussions*, 1–14.

Suh, A., & Prophet, J. (2018). The state of immersive technology research: A literature analysis. *Computers in Human Behaviour, 86*, 77–90.

Weibel, P. (2002). Narrated theory: Multiple projection and multiple narration. In M. Rieser & A. Zapp (Eds.), *New screen media: Cinema/art/narrative*. BFI & ZKM.

Willcock, S., Cooper, G., Addy, J., & Dearing, J. (2023). Earlier collapse of Anthropocene ecosystems driven by multiple faster and noisier drivers. *Nature Sustainability, 6*, 1331–1342.

Willemen, P. (2002). Reflections on digital imagery: Of mice and men. In M. Rieser & A. Zapp (Eds.), *New screen media: Cinema/art/narrative*. BFI & ZKM.

Chapter 3
Latest Advances and Challenges in Extreme Flood 3D Simulation

Ashish Sharma and Fiona M. Johnson

Abstract This chapter canvasses the latest developments in the modelling and communication of environmental extremes, with a focus on floods. Three scenarios are explored. The first refers to real-time prediction, including the current modelling basis that is adopted, and the visualisation/communication strategies in place. The second refers to an environmental extreme event that is conditional to a failure scenario, as is the case when an existing infrastructure (i.e. levee or spillway in an extreme flood) fails. The third, more complex scenario is the occurrence of a compound or joint extreme, possibly in the future, where extreme storms will intensify. A compound extreme here could represent a flood event that follows from an incident of rare storm conditions on a fire-damaged landscape. While the modelling challenges are significant, visualisation is even more challenging, as the scenario occurs under a hypothetical future. Demonstrating how coupled models can support the anticipation of extreme event scenarios, the chapter considers implications for risk assessment and communication that can support future preparedness and resilience. Surveying knowledge gaps that still need to be bridged, the authors formulate a list of key requirements in the fields of data availability, processing and representation.

Keywords Climate variability · Data integration · Environmental data · Flood modelling · Physics-based simulation · Real-time interactivity · Risk communication of unpredictability

A. Sharma (✉)
School of Civil & Environmental Engineering, The University of New South Wales, Kensington, NSW, Australia
e-mail: a.sharma@unsw.edu.au

F. M. Johnson
Water Research Centre, The University of New South Wales, Kensington, NSW, Australia
e-mail: f.johnson@unsw.edu.au

D. Del Favero et al. (eds.), *Climate Disaster Preparedness*, Arts, Research, Innovation and Society, https://doi.org/10.1007/978-3-031-56114-6_3

3.1 Introduction

Extremes, by definition, are very rare events. This makes them events that are difficult to anticipate or protect oneself from. They impact extensively, often over spatial and societal scales covering the entire countries, necessitating interventions that are instituted by governments responsible for ensuring society remains safe and its institutions operational. Flood extremes are a special type of naturally occurring extreme events that are modulated by the climate and also by the landscape they occur in. Wildfires are also modulated by both the climate and the landscape but evolve with greater randomness given their dynamic interaction with the elements over time. The occurrence of either floods or wildfires is a random extreme event characterisable by a suitable extreme value probability distribution. The evolution of such an extreme event, however, is deterministic. While both floods and wildfires are modulated by the same factors (climate and landscape), much of the discussion that follows focuses on floods, their prediction and the various factors that affect the extent of damages they result in.

Floods are the biggest and most severe natural disaster we face each year (Doocy et al., 2013). Apart from the loss of lives, floods disrupt daily activities, stop productivity and create a sense of helplessness that is socially and mentally disheartening. Technological advances have led to a doubling of life expectancy and per capita agricultural productivity in the last 100 years, yet flood deaths continue to rise (Doocy et al., 2013). The number of people affected will double within 20 years, due to rising populations, their concentration around floodplains and intensifying storm extremes resulting from planetary warming (Gassert et al., 2013).

Flood disasters are inevitably a result of an extreme storm event. However, if adequate flood protection infrastructure (such as reservoirs or levees) is in place, communities can remain largely unaffected. Enhanced atmospheric greenhouse gas concentrations and the increase in the moisture-holding capacity of a warmer atmosphere have resulted in a worldwide intensification of extreme storm events. As a result, extreme floods, especially over smaller catchments, are increasing and are expected to worsen with time (Sharma et al., 2018). Hence, the probability of failure for existing flood protection infrastructure is greater than when first designed. A consequence of this is an increasing frequency of flood disasters resulting from failure of existing flood infrastructure, such as failure of a dam upstream of a city, or rupture of a levee designed to keep flood waters at bay.

This chapter presents the rationale behind predicting flood impact in three different settings. First, we outline the factors that modulate a flood extreme, as well as the basis that is used to predict floods in advance. Next, we discuss a flood event that results in failure of the flood infrastructure, amplifying the damage caused, a scenario that is increasing in likelihood as a result of global warming. This is followed by a discussion of 'compound flood extremes', i.e. a flood event that is a result of not one but multiple extreme events occurring simultaneously. Such an event is difficult to plan for and is again increasing in likelihood as a result of global warming. We conclude by discussing how flood visualisation can assist with both real-time assistance and longer-term planning, along with the challenges that such visualisation systems typically face.

3.2 Real-Time Flood Warning Systems

So, how are flood warnings issued, and what are the key factors that introduce uncertainty into the quality of predictions made? To answer this question, one must first list the key inputs needed to quantify the evolution of a flood over a catchment. As mentioned before, the two main factors that modulate floods and make them different from one location to another are the contributing area (or landscape) and the climatic event that results in the flood. While the contributing area can be defined with ease given a topographic map and a catchment outlet, there exists uncertainty in how the climate interacts with this area to create the flood wave. A significantly greater uncertainty exists in the climate, which changes for the same catchment from one event to the other.

A key uncertainty in flood prediction is the specification of the model that could accept the climate as an input and estimate the flood magnitude as a function of time at the location of interest. The translation of the rainfall sequence to the flood is a deterministic process. Yet, for a model to be accurate, there needs to exist measured streamflow at the location of interest to ensure model parameters can be calibrated and the model thereby used into the future. When flood data is not available—as is the case in the vast majority of catchments, especially in low- and middle-income countries or in remote settings worldwide—the uncertainty in the flood prediction model becomes significant. There is extensive evidence that although the total population exposure to natural hazards in low- and middle-income countries is similar to high-income countries, then fatalities are far higher in the former (Strömberg, 2007; Lindersson et al., 2023). Flooding constrains population development opportunities, and 89% of the world's flood-exposed people live in low- and middle-income countries (Lindersson et al., 2023).

Another key uncertainty in flood prediction lies in the specification of the climate anomaly that translates into the flood. While the density of precipitation gauges is markedly greater than the corresponding streamflow measurements, these represent point precipitation and do not capture the significant variability across a catchment. Furthermore, precipitation gauges are few in remote areas, creating significant uncertainties in the depth and spatial distribution of recorded precipitation. Given floods are extreme events that are a result of extreme climatic anomalies, such observational uncertainties lead to large errors in ensuing flood warnings and predictions. When precipitation is not measured but predicted ahead of time, such uncertainties increase manifold (Ehrendorfer, 1997).

To address such uncertainties, modellers use surrogate data, representing both streamflow (to use directly for forecasting or to use as the basis for calibration) and the rainfall storm event (as an interpolated event based on point observations or using a weather forecasting approach that is based on satellite observations of storm

Fig. 3.1 The flood in Lismore (New South Wales, Australia) in March 2002. Shown are the streamflow measurement (Q, in m^3/s) at Leycester River at Rock Valley (Streamflow Station 203010) near Lismore and the surrogate runoff (SR) satellite-derived flow as per Yoon et al. (2022, 2023)

trajectories and the dynamics of storm creation). An example of a surrogate-streamflow model developed using publicly available satellite remote sensing data is presented in Fig. 3.1, with procedural details described by Yoon et al. (2022, 2023). No ground streamflow data was used in predicting the shown flood anomaly. Instead, the researchers used satellite retrievals that served as a surrogate to the actual streamflow over time. The implications of that for ungauged catchments, especially in remote settings, are significant.

However, observational-based predictions, such as those using remotely sensed data, generally do not provide enough time for action by communities and emergency managers. Thus, accurate forecasts of extremes are vital to increase the lead time available for preparation. Numerical weather prediction (NWP) models are vital components of flood warning systems. Despite the substantial advances in NWP skill over the last 40 years, which have assisted with improving event preparation, forecasts of precipitation are generally biased. Biases in precipitation are one of the most challenging aspects of NWP models to correct due to the presence of zero amounts (dry hours/days) as well as the highly skewed distribution of rainfall, with many small rainfall amounts and few very large or extreme rainfall events, which are the events that lead to floods. These errors in NWP rainfall simulations come from improperly resolved topography, temperature biases leading to convection biases and biases in the location of storm tracks. These issues cause underestimated high rainfall intensities, too much drizzle (Huang & Luo, 2017) and under-dispersed ensemble predictions (uncertainty ranges that are too confident). Most research into NWP rainfall correction has used quantile mapping (Hamill & Scheuerer, 2018). Quantile mapping corrects the modelled rainfall to match the observed rainfall across the full range of rainfall values from small to large, but it cannot correct incorrect sequencing of rainfall, e.g. how often a dry day is followed by a wet day, known as persistence biases. Bennett et al. (2014) addressed this problem by incorporating a Schaake shuffle, which retrospectively addresses the lack of

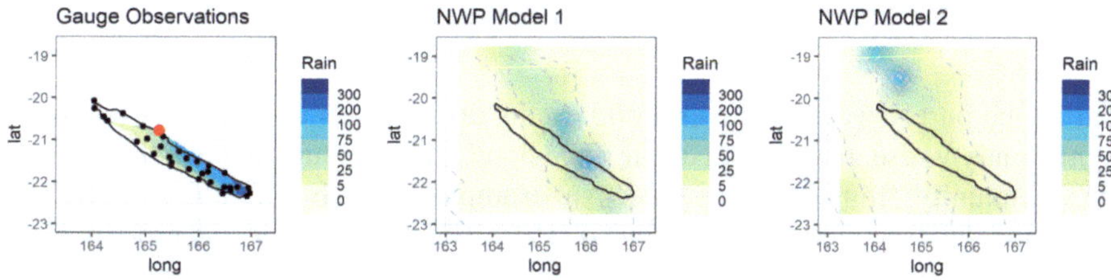

Fig. 3.2 Observed rainfall and forecast rainfall from models (in mm) for New Caledonia on 29 February 2020 for two NWP models at 1-day lead time showing large errors in both rainfall total and location

persistence. Recently, Jiang and Johnson (2023) proposed a new method for post-processing NWP forecasts using continuous wavelet transforms (CWT), coupled with quantile mapping. After decomposing the forecasts and observations using CWT, the amplitudes of the decomposed series are bias-corrected with quantile mapping, while the phases are randomised to correct timing errors. Spatial structure in the forecasts is improved by adopting the same phase randomisation at all gauge locations being corrected. A discrete wavelet-based approach combined with quantile mapping was used to bias-correct streamflow forecasts by Johnson (2023). It was found that calibrating the variance corrections over the full historical period provided marginally better forecast skill than a time-window-based approach (Fig. 3.2).

Pappenberger et al. (2015) quantified the monetary value of the European Flood Awareness System (EFAS) at 400:1 in terms of the potentially avoided flood damages across Europe compared to the cost of developing and operating EFAS. However, of the total Australian spending on disaster relief, 97% goes towards post-disaster recovery and only 3% to improved preparation and resilience (Productivity Commission, 2014), so there is a clear need to improve the focus on preparation for events. There have been calls for almost 20 years on the need for people-centred approaches in the design and operation of early warning systems (Basher, 2006), with the Sendai Framework (United Nations Office for Disaster Risk Reduction, 2015) continuing to emphasise the importance of people-centred preventative approaches to disaster risk. Traditionally, early warning systems have been viewed as a warning chain (Basher, 2006), a linear set of actions from observation to warning generation to dissemination. More recently, there has been recognition that instead of focusing on dissemination as the 'last mile' in the warning system, the system should be designed based on who the community is, their degree of vulnerability and their capacity (Marchezini, 2020). New and creative visualisation methods are thus urgently required to allow communities to better understand the threats for natural hazards and co-design effective warning systems to improve preparedness.

3.3 Flood Disasters Following Infrastructure Failure

Unlike the purely weather-driven flood events discussed in the previous section, a significantly worse disaster can occur if the flood infrastructure that exists to protect the community from flooding fails. An example of such a failure occurred in September 2023 in Derna, Libya (Saeed et al., 2023). This event was brought upon by anomalously warmer oceans and presence of high degrees of atmospheric moisture across the Mediterranean Sea and culminated in the collapse of two flood protection dams upstream of Derna. It resulted in a confirmed 4000 deaths with 10,000 missing, in a city with a total population of 90,000 residents.

Disasters such as that in Derna can be avoided if the flood protection infrastructure in place (here, the two dams that collapsed upstream of the city) operates as designed. In the design of such infrastructure, engineers typically derive an upper limit that a flood peak may assume, which is commonly referred to as the 'probable maximum flood' (PMF). The PMF is derived using an equivalent upper limit of the causative storm, referred to as the 'probable maximum precipitation' (PMP). Neither the PMF nor the PMP are absolute maximums—their estimation uses observations making the estimate a random variable with an extremely low probability of exceedance. Large dams around the world are often required to withstand a PMF, especially if communities live downstream.

The PMP is estimated using guidelines established by the World Meteorological Organization (WMO) and is proportional to the atmospheric moisture-carrying capacity of a warming atmosphere (2009). Figure 3.3 presents results from a recent study assessing change in the PMF over time, using both observational records on a surrogate of atmospheric moisture and climate model projections of the same surrogate up to the year 2100 (Visser et al., 2022). The change projected in the figure is surprisingly consistent across the landmass of Australia. Given the PMP represents an upper limit, it is prudent to assume that its projection for the year 2100 uses the more extreme SSP5–8.5 pathway, based on which one can expect the PMP will increase by 38% across existing reservoirs in Australia.

The projections of the increase in PMP by 38% in Fig. 3.3 implies that (a) PMFs will be increasing at least at the same rate as the PMP, (b) this increase will be remarkably consistent across climate zones and driven solely by a warming atmosphere and (c) the risk of failure for existing dams is now much greater than what it was when the dam was first constructed—with this risk increasing year after year as warming continues. While a PMF is by definition an extremely rare flood event that would never typically be recorded during the life span of a reservoir, global warming has made the PMF more likely. This means that disasters, such as happened at Derna, are becoming much more likely as we progress into this century.

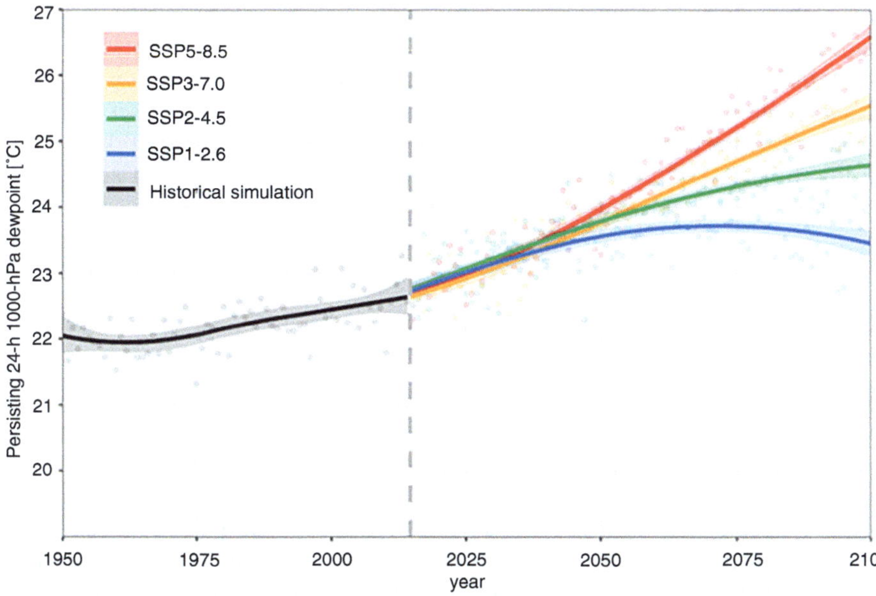

Fig. 3.3 Observed and climate model simulated persisting 24-h dew point temperatures up to the year 2100, under four plausible shared socioeconomic pathways (SSPs) (see Visser et al., 2022), the persisting 24-h dew point temperature being used by the WMO as a surrogate of the atmospheric moisture and a variable that is directly proportional to the PMP

3.4 Compound Flood Events

As outlined above, communities are safeguarded from the impact of floods by putting in place flood protection infrastructure, such as levees, dams or other barriers that stop or restrict the extent of damage that would otherwise occur. Such infrastructure is either designed on the basis of risk or designed to fail on average once in a certain number of years (Razavi et al., 2020). Typically, such design assumes the flood is a result of an extreme storm event that has the same risk of occurrence as the ensuing flood.

Of late, concerns have been growing that flood failure events are not a result of a single causative factor (i.e. an extreme storm) but an outcome that is due to two or more factors occurring simultaneously. An example of such an event may be two storm systems that occur simultaneously on two branches of a river system. This may result in extreme flood damage at their confluence, or the flooding of a coastal city that could be aggravated by extreme winds, resulting in a storm surge. Leonard et al. (2014) have presented a framework for assessing such compound events.

Moreover, there have been concerns that the risk of compound events is on the rise as a result of climate change. Two recent studies by Gu et al. (2022) and AghaKouchak et al. (2020) use observed data to establish that the risk of coastal floods, and the risk of high temperatures coupled with flood conditions, is rising across the world. How these pan out into the future is a question left largely untouched by the research community, although tools exist for simulating future compound extremes coupled to climate model simulations. These are downscaled to higher resolutions after correcting the simulations for systematic biases across all boundaries of the domain being downscaled (Kim et al., 2023). More work is needed on this front to better assess how society can be safeguarded from such future extremes as risks increase over time.

3.5 Visualisation of Flood Extremes

In preceding sections, we outlined three scenarios whereby flood disasters can impact communities downstream. The first scenario is the occurrence of a flood following an extreme storm event. The second scenario is a flood causing existing flood protection infrastructure to fail and result in magnified damage on communities downstream. The third scenario articulated the complexity in the flood generation mechanism, noting that floods may occur through a combination of causative events, such as a storm coupled with a coastal storm surge. All these scenarios can be represented through mathematical models of the climate, flood and landscape. Other data streams, such as satellite data or real-time rainfall and streamflow data, are also vital to understand the hazards. However, without effective visualisation of the data streams or model simulations, there is limited benefit in terms of facilitating preparedness. In this section, we review traditional and new developments in visualisation that can be leveraged for flood management and preparation.

Traditionally, flood visualisation has focused on mapping of flood extent, velocity and depth derived from hydrodynamic models. An excellent review on modelling and visualising flood inundation is provided by Teng et al. (2017). Flood mapping entails tracking the evolution of a flood wave in space and time as it follows the path of gravity and disperses over the downstream terrain. Such inundation modelling forms the basis for hazard mapping (a function of the maximum inundation depth), erosion modelling (based on flow velocity), flood insurance valuations (a function of the frequency and hazard) and a range of other actions. Flood inundation modelling utilises hydrodynamic modelling as the basis for tracking the evolution of the flood wave and offers a range of simplifications to reduce computational effort and produce inundation maps to varying degrees of accuracy. The resulting maps from these analyses are generally presented as part of a flood study, are static and represent averaged long-term risk of floods at any particular location.

Increasingly though, a range of new approaches are being investigated to better communicate the risk and impacts of flood events. Participatory practices for flood mapping are increasingly being considered (e.g. Auliagisni et al., 2022). For example, Disaster Relief Australia (2023) has recently implemented 'Big Map Workshops' to increase community resilience to flood events in order to improve post-event recovery for both floods and fires. Instead of the flood map being part of a report, it is a room-sized visualisation of the community and its risk. Such workshops allow a range of stakeholders to better understand the dynamic nature of such natural hazards and thereby prepare effectively as a community for the next event.

Similarly, Li et al. (2022) describe a three-dimensional storytelling approach to visualise the impacts of infrastructure failure, such as a dam-break flood discussed earlier in this chapter. Storytelling approaches have been used more widely in communicating climate change adaptation options but are yet to be routinely used in floodplain management. Such methods provide exciting promise for helping communities to understand the dynamic nature of flood events compared to the static flood maps discussed above.

Other methods have been developed using strength-based approaches for engaging stakeholders. For example, relatively simple methods such as schematic diagrams can be used to provide real-time updates from a range of data streams and forecasts. The advantage of these schematic diagrams is that they do not rely on high spatial awareness skills for users. Yasmin et al. (2023) point to the importance of such inclusive practices in the design of communications. Similarly, Auliagisni et al. (2022) argue for the importance of engagement with a wide range of stakeholders (Bakhtiari et al., 2024), such as insurance providers and policy makers, in the use of new visualisation approaches such as digital twins or augmented and virtual reality. One of the major open questions in visualisation approaches is how to usefully communicate uncertainty in predictions and forecasts. This is vital for improving the ability of emergency services to respond to increasing threats from compound events, which will stretch already limited resources.

3.6 Conclusion

The World Resources Institute publishes current and projected flood damages on a country-wide basis (Gassert et al., 2013). According to this, Australia incurs an annual flood-related damage of US$18.7 billion, a number that will increase to US$29.9 billion by 2030. This increase is small when compared to Australia's neighbours in South-East Asia, with three-fold (Malaysia) to 15-fold (Myanmar, Cambodia) increases predicted to severely impact resident societies and economies. Much of this damage relates to urban flooding, which is expected to increase most significantly due to intensification of storms and lack of the significant antecedent storage, which dampens flow peaks in rural settings. Much of such damage can be contained if adequate flood protection infrastructure is put in place. However, the damage increases when existing infrastructure becomes inadequate, which we have argued in this chapter. This is indeed occurring as the world records higher temperatures each year in the recent past.

We outlined three scenarios that are important to study as the world prepares for flood extremes on a different scale. A significantly increased exposure to risk exists as populations concentrate in urban centres that often lie near a major water body that serves as the source of flood water supply. The first of these scenarios represented flood events that are caused solely by extreme storms and argued that more could be done to predict such events in advance through a mixture of modelling and weather forecasting. The second and third flood scenarios referred to design extremes that represent hypothetical events used to design flood protection infrastructure that is increasingly vulnerable under condition of climate change. In all of the scenarios discussed, there is a common need of methodologies to visualise flood damage, to assist in planning, risk/hazard quantification as well as devising methods for evacuation, all serving as important legs of a comprehensive damage control strategy. Such a strategy is sorely needed as flood damages rise with time (Wasko et al., 2021), a rise that will only get worse as temperatures increase and populations converge on concentrated urban centres across the world. The capability to viscerally preview the dynamic evolution and likely extent of future extreme flooding events can help protect and save lives as well as safeguard assets and infrastructure by enabling us to build effective preparedness for a hotter future that is inevitably coming to meet us.

References

AghaKouchak, A., Chiang, F., Huning, L. S., et al. (2020). Climate extremes and compound hazards in a warming world. *Annual Review of Earth and Planetary Sciences, 48*, 519–548.

Auliagisni, W., Wilkinson, S., & Elkharboutly, M. (2022). Using community-based flood maps to explain flood hazards in Northland, New Zealand. *Progress in Disaster Science, 14*, 100229.

Bakhtiari, V., Piadeh, F., Chen, A. S., & Behzadian, K. (2024). Stakeholder analysis in the application of cutting-edge digital visualisation technologies for urban flood risk management: A critical review. *Expert Systems with Applications, 236*, 121426.

Basher, R. (2006). Global early warning systems for natural hazards: systematic and people-centred. *Philosophical Transactions of the Royal Society A – Mathematical Physical and Engineering Sciences, 364*(1845), 2167–2182.

Bennett, J. C., Robertson, D. E., Shrestha, D. L., Wang, Q. J., Enever, D., Hapuarachchi, P., & Tuteja, N. K. (2014). A system for continuous hydrological ensemble forecasting (SCHEF) to lead times of 9 days. *Journal of Hydrology, 519*, 2832–2846.

Disaster Relief Australia. (2023). Big map. *Disasterreliefaus.org*. https://shorturl.at/jvKU2. Accessed 13 Dec 2023.

Doocy, S., Daniels, A., Murray, S., & Kirsch, T. (2013). The human impact of floods: A historical review of events 1980–2009 and systematic literature review. *PLOS Currents: Disasters, 1*.

Ehrendorfer, M. (1997). Predicting the uncertainty of numerical weather forecasts: A review. *Meteorologische Zeitschrift – Berlin, 6*, 147–183.

Gassert, F., Luck, M., Landis, M., Reig, P., & Shiao, T. (2013). *Aqueduct global maps 2.0*. World Resources Institute.

Gu, L., Chen, J., Yin, J., Slater, L., Wang, H.-M., Guo, Q., Feng, M., Qin, H., & Zhao, T. (2022). Global increases in compound flood-hot extreme hazards under climate warming. *Geophysical Research Letters, 49*(8), e2022GL097726.

Hamill, T. M., & Scheuerer, M. (2018). Probabilistic precipitation forecast postprocessing using quantile mapping and rank-weighted best-member dressing. *Monthly Weather Review, 146*(12), 4079–4098.

Huang, L., & Luo, Y. (2017). Evaluation of quantitative precipitation forecasts by TIGGE ensembles for south China during the presummer rainy season. *Journal of Geophysical Research: Atmospheres, 122*(16), 8494–8516.

Jiang, Z., & Johnson, F. M. (2023). A new method for postprocessing numerical weather predictions using quantile mapping in the frequency domain. *Monthly Weather Review, 151*(8), 1909–1925.

Johnson, F. M. (2023). How useful are global streamflow forecasts for Australian catchments? In *Hydrology and water resources symposium*. Engineers Australia.

Kim, Y., Evans, J. P., & Sharma, A. (2023). Correcting systematic biases in regional climate model boundary variables for improved simulation of high-impact compound events. *iScience, 26*(9), 107696.

Leonard, M., Westra, S., Phatak, A., et al. (2014). A compound event framework for understanding extreme impacts. *Wiley Interdisciplinary Reviews: Climate Change, 5*(1), 113–128.

Li, W., Zhu, J., Haunert, J. H., Fu, L., Zhu, Q., & Dehbi, Y. (2022). Three-dimensional virtual representation for the whole process of dam-break floods from a geospatial storytelling perspective. *International Journal of Digital Earth, 15*(1), 1637–1656.

Lindersson, S., Raffetti, E., Rusca, M., Brandimarte, L., Mård, J., & Di Baldassarre, G. (2023). The wider the gap between rich and poor the higher the flood mortality. *Nature Sustainability, 6*(8), 995–1005.

Marchezini, V. (2020). "What is a sociologist doing here?" An unconventional people-centered approach to improve warning implementation in the Sendai framework for disaster risk reduction. *International Journal of Disaster Risk Science, 11*(2), 218–229.

Pappenberger, F., Cloke, H. L., Parker, D. J., Wetterhall, F., Richardson, D. S., & Thielen, J. (2015). The monetary benefit of early flood warnings in Europe. *Environmental Science & Policy, 51*, 278–291.

Productivity Commission. (2014). *Natural disaster funding arrangements (Inquiry report no. 74)*. Canberra. JEL code: H77, H84.

Razavi, S., Gober, P., Maier, H. R., Brouwer, R., & Wheater, H. (2020). Anthropocene flooding: Challenges for science and society. *Hydrological Processes, 34*, 1996–2000.

Saeed, A., Ibrahim, N., & Mahmoud, Y. (2023). Libya floods: The flawed response that increased Derna death toll. *BBC News*. https://shorturl.at/fwV49. Accessed 13 Dec 2023.

Sharma, A., Wasko, C., & Lettenmaier, D. P. (2018). If precipitation extremes are increasing, why aren't floods? *Water Resources Research, 54*(11), 8545–8551.

Strömberg, D. (2007). Natural disasters, economic development, and humanitarian aid. *Journal of Economic Perspectives, 21*(3), 199–222.

Teng, J., Jakeman, A. J., Vaze, J., Croke, B. F. W., Dutta, D., & Kim, S. (2017). Flood inundation modelling: A review of methods, recent advances and uncertainty analysis. *Environmental Modelling & Software, 90*, 201–216.

United Nations Office for Disaster Risk Reduction. (2015). *Sendai framework for disaster risk reduction 2015–2030*. United Nations.

Visser, J. B., Kim, S., Wasko, C., Nathan, R., & Sharma, A. (2022). The impact of climate change on operational probable maximum precipitation estimates. *Water Resources Research, 58*, 11.

Wasko, C., Sharma, A., & Pui, A. (2021). Linking temperature to catastrophe damages from hydrologic and meteorological extremes. *Journal of Hydrology, 602*, 126731.

World Meteorological Organization. (2009). *Manual on estimation of probable maximum precipitation (PMP)*. WMO.

Yasmin, T., Khamis, K., Ross, A., et al. (2023). Inclusiveness in designing an early warning system for flood resilience. *Natural Hazards and Earth System Sciences, 23*(2), 667–674.

Yoon, H. N., Marshall, L., Sharma, A., & Kim, S. (2022). Bayesian model calibration using surrogate streamflow in ungauged catchments. *Water Resources Research, 58*, 1.

Yoon, H. N., Marshall, L., & Sharma, A. (2023). Beyond river discharge gauging: hydrologic predictions using remote sensing alone. *Environmental Research Letters, 18*(3), 034015.

Chapter 4
Intelligent Architectures for Extreme Event Visualisation

Yang Song, Maurice Pagnucco, Frank Wu, Ali Asadipour, and Michael J. Ostwald

Abstract Realistic immersive visualisation can provide a valuable method for studying extreme events and enhancing our understanding of their complexity, underlying dynamics and human impacts. However, existing approaches are often limited by their lack of scalability and incapacity to adapt to diverse scenarios. In this chapter, we present a review of existing methodologies in intelligent visualisation of extreme events, focusing on physical modelling, learning-based simulation and graphic visualisation. We then suggest that various methodologies based on deep learning and, particularly, generative artificial intelligence (AI) can be incorporated into this domain to produce more effective outcomes. Using generative AI, extreme events can be simulated, combining past data with support for users to manipulate a range of environmental factors. This approach enables realistic simulation of diverse hypothetical scenarios. In parallel, generative AI methods can be developed for graphic visualisation components to enhance the efficiency of the system. The integration of generative AI with extreme event modelling presents an exciting opportunity for the research community to rapidly develop a deeper understanding of extreme events, as well as the corresponding preparedness, response and management strategies.

Y. Song (✉) · M. Pagnucco
School of Computer Science & Engineering, The University of New South Wales (UNSW), Kensington, NSW, Australia
e-mail: yang.song1@unsw.edu.au; morri@unsw.edu.au

F. Wu
iCinema Centre, The University of New South Wales (UNSW), Kensington, NSW, Australia
e-mail: frank.wu1@unsw.edu.au

A. Asadipour
Computer Science Research Centre, Royal College of Art, London, UK
e-mail: ali.asadipour@rca.ac.uk

M. J. Ostwald
School of Built Environment, The University of New South Wales, Kensington, NSW, Australia
e-mail: m.ostwald@unsw.edu.au

D. Del Favero et al. (eds.), *Climate Disaster Preparedness*, Arts, Research, Innovation and Society, https://doi.org/10.1007/978-3-031-56114-6_4

Keywords Intelligent visualisation · Dynamics · Modelling and simulation · Generative AI · Deep learning

4.1 Introduction

Extreme events such as earthquakes, floods and wildfires have a significant impact on both the natural environment and human society. To effectively predict, prepare for and manage the impact of extreme events, researchers have developed a range of physics-based modelling methodologies to understand the underlying dynamics of such events. When these modelling methods are integrated into an immersive visual environment, researchers and domain experts can interact with systems and better understand the complex nature of extreme events and human responses. This increased understanding relies on three factors. The first is the physical presence users feel in immersive environments, and the second is how this type of presence heightens intuitive understanding and spatial cognition. The third factor is associated with a capacity to interact with or shape the environment. In an immersive visualisation, users can specify key environmental factors that would affect the dynamics of extreme events, and the system will then adapt the visualisation accordingly to provide a highly naturalistic depiction of various scenarios. Such *intelligent visualisation* systems integrating physics- and data-driven modelling and simulation will be highly effective for preparing communities, designing response strategies and training first responders.

Currently, the simulation of earthquakes with supercomputers has been an active research field, and there is significant effort being invested by researchers in developing open-access datasets to facilitate further data-driven research (Kovner, 2022). There has also been significant research for fire and flood modelling using both physics-based and machine learning approaches (Jain et al., 2020; Teng et al., 2017). However, there is relatively less research specifically focused on immersive visualisation for extreme events, especially for intelligent visualisation that can adapt dynamically to different environments in simulated scenarios.

In this chapter, we will first provide a review of representative approaches that build towards *intelligent visualisation* of extreme events. We consider that intelligent visualisation is a computational pipeline that consists of (i) *modelling*, (ii) *simulation* and (iii) *graphic visualisation*. While modelling and simulation focus on data generation, graphic visualisation uses computer graphics algorithms to represent the generated data in a visually immersive and realistic way. Next, motivated by the recent success of deep learning and generative artificial intelligence (AI), we will present suggestions for how *generative AI methodologies* can be incorporated into the visualisation of extreme events. Finally, we will discuss how different generative AI methods can support the various components required in a visualisation pipeline for extreme events.

4.2 Intelligent Visualisation of Extreme Events

While the noun "visualisation" often refers to the graphic presentation or representation of image-based data, in the present context, we focus on *intelligent visualisation*, which consists of a complete computational pipeline including modelling, simulation and graphic visualisation (Fig. 4.1). Such an intelligent visualisation system will be able to generate data representations of extreme events based on physical modelling or learning-based modelling and simulations, which are then visualised in high resolution with support for user interaction and immersive experiences. This section discusses examples of each of the three pipeline stages.

4.2.1 Physical Modelling

The objective of using physical modelling for extreme events is to develop mathematical models that replicate the underlying principles and behaviours of the dynamic evolution of these events. For instance, through physical modelling, studies have investigated the effect of wind, slope, fuel moisture, fuel structure and ignition setting on the rate of spread and intensity of bushfires (Sharples & Hilton, 2020).

Fire modelling approaches have evolved from initial one-dimensional (1D) rate of spread (RoS) estimations to the more intricate 2D or 3D simulations that depict the expansion of fire perimeters in spatial contexts. Physical fire models follow the same fundamental principles of physics but differ in choosing the governing equations and implementations. They also vary in complexity and dimensionality. For example, the classical approach for fire spread modelling (Weber, 1991) was initially a 1D model to predict RoS based on the flux of energy and later extended to a 2D plane. A model called WFDS was later developed for 3D simulations to resolve different physical process stages (Mell et al., 2007).

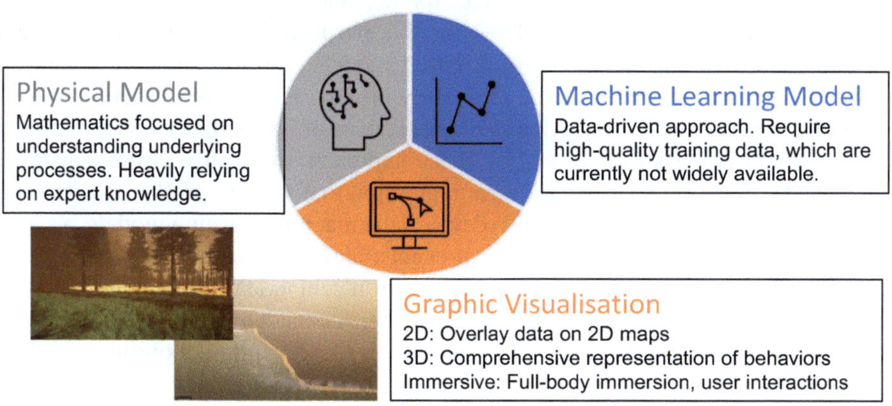

Fig. 4.1 Overview of intelligent visualisation

Some physical models have been integrated into software packages for various application domains. For instance, WRF-SFIRE (Mandel et al., 2014) provides a coupled meteorological and fire spread model. This integrated system accounts for dynamic interactions between weather conditions and fire behaviour by solving intricate physical and chemical processes within a high-resolution 3D grid-like domain. With its capability for advanced solution modelling, WRF-SFIRE has been widely used by researchers in fire dynamics. FlamMap (Finney, 2006) "is a fire analysis desktop application that [...] [includes a suite of functions that can simulate] potential fire behaviour characteristics, [such as] fire growth and spread, and conditional burn probabilities under constant environmental conditions" (Finney, 2023). It also encapsulates FARSITE (Finney, 2004), which computes wildfire growth and behaviour for longer time periods under heterogeneous conditions.

In practice, physical models are primarily adopted in behaviour analysis of fire and flood events rather than operational use, mainly due to the challenges associated with validation and computational demands. Moreover, employing physical models necessitates meticulous manual input that requires domain expertise. This input encompasses aspects such as defining initial geometry and domain parameters, specifying fire source characteristics and configuring simulation parameters, among others. These inputs cannot be accurately determined without a rigorous and deep understanding of the underlying physics involved. Consequently, physical models tend to present a steep learning curve for researchers who do not have a background in these disciplines.

4.2.2 Learning-Based Modelling and Simulation

Because the principles of physical modelling are founded in expert knowledge, its capability for modelling complex or new scenarios is also inherently limited by experts' existing knowledge. To overcome this situation, recent approaches have explored the use of learning-based methodologies to reveal previously hidden patterns in historical or experimental data for fire and flood behaviour analysis. A diverse range of methods is available for this purpose, including both statistical machine learning and neural network-based models, and the choice of methods typically depends on the available scalability of data. For instance, there has been extensive research on flooding due to intensification of heavy rainfall under climate change conditions (Ho et al., 2023). These approaches typically utilise statistical machine learning models such as linear regression to discover correlations between events and environmental factors. To develop the machine learning models, a set of data would be collected from historical events, and various data-driven computations are applied to it, before it is fit for use in machine learning models. Similar approaches have also been developed using experimental data to address the limitations of historical data. For example, data samples from outdoor experimental fires and natural, more intense wildfires can be obtained using logistic and non-linear regression models for the rate of fire spread (Cruz et al., 2021). Such models demonstrate an ability to represent a broad range of wildfire behaviour adaptions to the effects of wind speed, fuel structure and various landscape conditions.

Other types of statistical machine learning models have also been incorporated into extreme event modelling and simulation. In an example of the application of RoS estimation, a Bayesian model was developed based on weather variables so that the model can effectively accommodate the variability in model inputs and uncertainty associated with RoS prediction (Storey et al., 2021). Bayesian models have also been adapted in flood modelling to estimate the frequency of extreme flood events based on historical records (Parkes & Demeritt, 2016).

With the recent development of neural networks and, particularly, deep learning, some approaches have been developed to perform modelling or simulation based on higher-dimensional data, such as satellite images, for fire spread modelling or simulation. For example, the FireCast method (Radke et al., 2019) is a simple convolutional neural network model developed based on satellite images and weather data. To address the issue of limited training data, weather interpolation and data augmentation techniques are employed. A similar approach was developed (Yang et al., 2021), incorporating "ground truth labels" obtained from public datasets. In another recent study, various data sources, "including topography, weather [conditions], [...], vegetation, and population density" (Huot et al., 2022) as well as satellite images, are combined to create a comprehensive dataset for predicting next-day wildfire spread. Huot et al. then formulate the prediction "as an image segmentation [task to] classify each area as either containing fire [or not], given the location [of] the fire of the previous day". A convolutional autoencoder is developed for the segmentation and demonstrates higher performance than other machine learning approaches based on logistic regression and random forest algorithms. In another approach, Hodges et al. (2019) consider the challenge of collecting sufficient amounts of training data to support a robust machine learning process. In response, they generate synthetic data using Rothermel (Scott & Burgan, 2005) for homogeneous landscapes and FARSITE for heterogeneous spreads. A deep convolutional inverse graphics network is then developed using the synthetic data to predict fire spread.

While learning-based approaches like these can overcome the problem of limited domain knowledge and represent a more diverse data distribution, the capabilities of existing approaches are still limited. Current applications of learning-based approaches are mainly focused on predicting the frequency of fire or flood events and are typically formulated as regression, classification or segmentation problems. However, such methods are not designed for generating realistic simulations of the dynamic behaviours of extreme events in hypothetical scenarios with diverse environmental conditions, especially when user interaction and adaptive simulation are expected. Moreover, the performance of machine learning models is highly dependent on large-scale, high-quality training data. While researchers have devoted substantial time to creating open-access datasets, these are still quite small scale, which then limits the performance and generalisability of the developed systems. We expect to see more developments in dataset creation and integration of learning-based approaches with knowledge-driven physical modelling, which would effectively address this issue.

4.2.3 Graphic Visualisation

The graphic visualisation component refers to the stage where real, modelling or simulation data is rendered and displayed in 2D or 3D. Typically, 2D visualisation is conducted by overlaying data on maps or displaying it as plots, whereas 3D visualisation provides a more comprehensive and realistic representation of an event's behaviour that is often rendered over a 3D map. A comprehensive survey of visualisation systems for wildfires has recently been published (Tirado Cortes et al., 2023). Here, we provide an overview of 3D methods in graphic visualisation that have been utilised in the domain of extreme events since 3D methods require more complex processing steps.

A representative system of 3D wildfire visualisation is presented by Castrillon et al. (2011), where FARSITE is used to generate the data to be visualised, such as fire perimeters, the intensity of flames and velocity of the fire front. A graphical interface is then built in a 3D Multiplayer Geographical Environment, integrating geographical layers and 3D objects over virtual terrain. The module for fire visualisation is developed based on two particle systems for modelling flame and smoke, which are controlled by an emitter specifying the behaviours of particles. The propagation of fire is also modelled by curve morphing techniques to update the perimeters of fires and generate animations. Various optimisation techniques are also implemented to adaptively reduce the mesh vertices and number of particles so that the visualisation can be realistic while reducing the graphic complexity. There are also examples of systems that introduce more user interaction functionalities to update the rendering and visualisation. For instance, in one system (Wahlqvist et al., 2021), users can change the views, data inputs and timestamps, and the visualisation can give valuable insights into the effect of fire spread on population areas.

Overall, while advanced graphics techniques can be implemented to achieve highly realistic visualisation of extreme events, significant advances are needed to support computational modelling for specific event scenarios. There is very limited support for dynamically updating the rendering of different scenes. Changing of environments will thus require extensive effort redesigning the underlying 3D models. 3D computer graphics also require extensive computational resources. To enhance efficiency, current approaches often resort to approximation algorithms (Byari et al., 2022) that reduce the spatial resolution or realism of the visualisation. While recent advancements in deep learning and generative AI have demonstrated impressive progress in computer graphics (Lefohn, 2023), more work is needed for adapting such methods for visualisation of extreme events.

4.3 Generative AI in Visualisation

While generative AI has attracted considerable public attention due to the popularity of ChatGPT, we believe the development of generative adversarial networks (GANs) (Goodfellow et al., 2014) marks the start of generative AI for images. By training a simple deep learning model with a generator and a discriminator, GANs can generate new images that resemble the original imaging domain. Many improvements to the original GAN structure have since been developed for various objectives, such as style transfer, image super-resolution and image editing, leading to diverse applications. Conversely, GANs can be difficult to train, and the generated images often lack diversity. Hence, other generative AI models are proposed, such as variational autoencoders (VAEs) (Kingma & Welling, 2019) and diffusion models (DMs) (Ho et al., 2020), although VAEs tend to produce images with lower quality and DMs can be slow when generating images. More recently, deep learning models have been adapted into computer graphics, such as neural radiance field (NeRF) (Mildenhall et al., 2020) and its variants, achieving both efficient and realistic graphic rendering and visualisation. Nevertheless, while significant research and industry development have been conducted on generative AI, relatively little work has been done specifically for the visualisation of extreme events. Here, we describe some representative studies of generative AI models in related application domains, which are useful precedents for adapting generative AI for visualisation of extreme events.

4.3.1 Image Generation

A typical GAN model contains two components: a generator that creates synthesised images and a discriminator that distinguishes between real and generated images. During the training process, the aim is to derive a generator that can create highly realistic images so that the discriminator cannot separate them from the real images. As a result, the trained GAN generator can be used to create new, high-quality images during the inference process. Many variants of the standard GAN model have been proposed, some customised for specific applications, while others address fundamental limitations in the GAN model, such as the difficulty of training and problems with mode collapse. A recent survey paper (Wang et al., 2021) presents a comprehensive overview of this field.

One example of the use of GANs in extreme event visualisation generates photo-realistic images showing how floods can affect the environment (Schmidt et al., 2022). The approach, named ClimateGAN, can generate flooded scenes with 1-metre flood levels based on arbitrary street-level scenes such as Google Street View images. The model consists of two modules: a Masker module for predicting the image regions that should be under water and a GAN-based Painter module to generate water textures based on the Masker's prediction. To train the model, paired images of before and after flooding would be needed, which are, however, rare and cannot be easily collected. Therefore, in ClimateGAN, a virtual world is created using the Unity3D engine to simulate urban, suburban and rural environments, which are then flooded with 1m of water to generate the paired training data. A smaller dataset of real images was also collected to enhance the training of the model. While ClimateGAN generates realistic images, it is difficult to extend it to floods of different heights, mainly due to the difficulty of data collection.

In contrast to GANs, DMs are inspired by non-equilibrium thermodynamics. A DM consists of two "processes, the forward diffusion process [which] defines a [Markov] chain of diffusion steps to slowly add random noise to data, [and] the reverse diffusion process" (Niu et al., 2023), which learns to reverse the forward process to construct desired data outputs from the noise. While DMs are typically much slower than GANs during the generation process, various techniques have been developed to enhance their efficiency. DMs have thus gained significant interest in the research community and industry, creating popular tools such as DALLE 2, because of their exceptional capabilities in creating high-quality, realistic and diverse images.

DMs have also been used for weather forecasting. For example, a recent approach (Chen et al., 2023), named SwinRDM, performs weather forecasting via a variational recurrent neural network and then interpolates the forecasting output via a diffusion-based super-resolution module. As a result, SwinRDM can provide global weather forecasting at 0.25-degree resolution without incurring an excessively high computational cost.

Based on these precedents, similar models can be developed for the visualisation of extreme events, such as using GANs or DMs to generate scenes showing wildfire spread or changes in landscapes after an extreme event episode. Similar to ClimateGAN, the difficulty would lie in data collection, as generative AI models require large-scale training data. Images of certain view angles would be easier, such as aerial images from satellite imagery. In other cases, a simulated environment might be the best approach to generate a sufficient amount of training data. Moreover, for extreme events, the realism of generated images is critical, and they need to adjust to different environmental conditions. To achieve this, it is possible to integrate text prompts in the image generation process via DMs or image templates as conditional input for GANs. Such information can be used to explicitly guide the image generation process so that the outputs can better approximate the expected scenarios following certain environmental variables.

4.3.2 Dynamic Simulation

The dynamic evolution of an extreme event, reflected in the rapid motion (direction and velocity) of the fire or flood, is an important aspect that cannot simply be represented by a sequence of images. For example, while the spread of a wildfire recorded in satellite imagery might be viewed once every few hours, a 3D visualisation of fire events in the first-person immersive view would require real-time dynamic update, and the information about evolution and motion also conveys causal effects according to the environmental conditions. While generative AI models have demonstrated impressive performance for single images, relatively fewer research studies have been conducted on generating dynamic time-lapse data or videos, often due to the significant requirement for computational power.

In a recent study (Chu et al., 2021), a GAN-based deep learning model is developed for fluid simulation, where the fluid can morph dynamically depending on several "control modalities, including obstacles, physical parameters, kinetic energy and vorticity" (ibid.). Interestingly, the model "explicitly embeds physical quantities into the learned latent space" (ibid.) so that the control parameters can effectively impact the generation of output and enhance their diversity. To train the model, a training dataset is created via simulation to generate pairs of images representing the density and velocity information. The dataset also introduces samples showing different velocities of moving obstacles so that the model training can be exposed to a variety of cases. The evaluation results show that the approach delivers higher performance than the other more standard GAN- or VAE-based models. However, as with all generative AI methods and especially GAN models, the simulation outputs do not always adapt well to user controls.

For dynamic or video generation, unlike the above-mentioned approach that involves explicit physical modelling, most methods choose to incorporate motion information via more traditional computer vision algorithms, such as optical flow. For instance, with DTVNet (Zhang et al., 2020), the generation framework takes in a single landscape image and then generates diversified time-lapse videos based on normalised motion vectors. The network contains two modules: an optical flow encoder that estimates the optical flow between consecutive images and a dynamic video generator that follows a GAN-based architecture and constructs the video frames by learning motion and content information. DTVNet experiments were performed on a dataset "containing dynamic sky scenes, [including a] cloudy sky with moving clouds and [...] a starry sky with moving stars" (Xiong et al., 2018). Evaluation of DTVNet in human user studies shows improved performance over other GAN models.

DMs have also been applied to dynamic scene or video generation, which typically shows more impressive results than GAN-based models but requires text prompts. For example, Imagen Video (Ho et al., 2022) is a text-conditional video generation system. Compared to other approaches, Imagen Video achieves high-definition video generation producing videos of 128 frames of 1280 × 768 pixels at 24 frames per second. It achieves this with a cascade of DMs containing a sequence of spatial and temporal super-resolution processes. While the method achieves remarkable performance, it was trained on an "internal dataset [containing] 14 million video–text pairs and 60 million image–text pairs" (Ho et al., 2022) as well as other large-scale public datasets. While this requirement can be prohibitive, for developing a domain-specific model, such as for simulating wildfires, a much smaller dataset should be feasible to achieve promising results.

Currently, DMs have demonstrated impressive performance for video generation, which can be a possible approach for generating dynamic simulation of extreme events. Customisations of the models would be required so that environmental variables can be effectively integrated in place of the text prompts. GANs, on the other hand, can be more flexible in terms of introducing environmental variables into the model. However, domain-specific customisation will also be required especially to encourage the diversity of data generation. Overall, in a manner similar to generative AI for image generation, a key design consideration would be the dataset. Large-scale datasets that can closely represent real data distribution and diversity would be valuable for developing such models. To accommodate the limitations of datasets, other techniques would need to be exploited, such as introducing explicit physical modelling, performing advanced data augmentation or integrating pretrained models with transfer learning.

4.4 Conclusion

Extreme climate events such as floods and wildfires pose a particular challenge to society. To better prepare the community and first responders for such unpredictable events, new methods are required. Intelligent visualisation facilitates (i) picturing diverse scenarios, (ii) developing rich and dynamic narratives from them, (iii) communicating the threats they entail and (iv) supporting people to rehearse their responses to these threats. As such, intelligent visualisation is central to both gaining new insights into extreme events and translating this knowledge to stakeholders.

Moreover, human perception of the environment and its ability to adapt to dynamic changes depend on the rapid acquisition of real-time sensory data for processing and swift response. The remarkable capacity of the brain to efficiently allocate computational resources and direct pertinent data streams to the relevant cortical regions for planning somatosensory reactions empowers us to manage these fluctuations. Nevertheless, our capacity to make well-informed decisions and respond appropriately to unfamiliar situations (dealing with uncertainty) remains significantly underexplored. Acknowledging that immersion is a multifaceted experience intricately influenced by various sensory modalities, this chapter places its primary emphasis on the saliency of visual information as a key driver of information acquisition when extreme events occur.

In this chapter, we provided a review of current methodologies in intelligent visualisation, focusing on physical modelling, learning-based modelling and simulation as well as graphic visualisation components. Then, considering the widespread success of deep learning and particularly generative AI models, we hypothesised that such models can also be adapted for the visualisation of extreme events. We thus presented several representative generative AI approaches in related application domains and discussed various design considerations when developing such approaches for extreme event visualisation. Ultimately, this chapter can be viewed as both a review and position paper for the emerging topic of intelligent visualisation for extreme events.

References

Byari, M., Bernoussi, A., Jellouli, O., Ouardouz, M., & Amharref, M. (2022). Multi-scale 3D cellular automata modelling. *Chaos, Solutions & Fractals, 164*, 112653.

Castrillon, M., Jorge, P., Lopez, I., Macias, A., et al. (2011). Forecasting and visualization of wildfires in a 3D geographical information system. *Computers & Geosciences, 37*(3), 390–396.

Chen, L., Du, F., Hu, Y., Wang, Z., & Wang, F. (2023). SwinRDM: Integrate SwinRNN with diffusion model towards high-resolution and high-quality weather forecasting. In Y. Chen & J. Neville (Eds.), *AAAI conference on AI* (pp. 322–330). AAAI.

Chu, M., Thuerey, N., Seidel, H., Theobalt, C., & Zayer, R. (2021). Learning meaningful controls for fluids. *ACM Transactions on Graphics, 40*(4), 100.

Cruz, M., Cheney, N., Gould, J., McCaw, W., et al. (2021). An empirical-based model for predicting the forward spread rate of wildfires in eucalypt forests. *International Journal of Wildland Fire, 31*(1), 81–95.

Finney, M. (2004). *FARSITE: Fire area simulator—Model development an evaluation* (Research paper). US Department of Agriculture & Forest Service

Finney, M. (2006). An overview of FlamMap fire modelling capabilities. In P. L. Andrews & B. W. Butler (Eds.), *Fuels management—How to measure success* (pp. 213–220). US Department of Agriculture & Forest Service.

Finney, M. (2023). FamMap. *Missoula Fire Sciences Laboratory*. https://t1p.de/htwci. Accessed 17 Dec 2023.

Goodfellow, I., Pouget-Abadie, J., Mirza, M., Xu, B., et al. (2014). Generative adversarial nets. In Z. Ghahramani, M. Welling, C. Cortes, N. Lawrence, & K. Q. Weinberger (Eds.), *International conference on neural information processing systems* (pp. 2672–2680). NeuIPS.

Ho, J., Jain, A., & Abbeel, P. (2020). Denoising diffusion probabilistic models. In H. Larochelle, M. Ranzato, R. Hadsell, M. F. Balcan, & H. Lin (Eds.), *International conference on neural information processing systems* (pp. 1–20). NeuIPS.

Ho, J., Chan, W., Saharia, C., Whang, J., et al. (2022). Imagen Video: High-definition video generation with diffusion models. *arXiv*: 2210.02303.

Ho, M., Wasko, C., O'Shea, D., Nathan, R., et al. (2023). Changes in flood-associated rainfall losses under climate change. *Journal of Hydrology, 625*, 129950.

Hodges, J., Lattimer, B., & Hughes, J. (2019). Wildland fire spread modelling using convolutional neural networks. *Fire Technology, 55*, 2115–2142.

Huot, F., Hu, R., Goyal, N., Sankar, T., et al. (2022). Next day wildfire spread: A ML dataset to predict wildfire spreading from remote-sensing data. *IEEE Transactions on Geoscience and Remote Sensing, 60*, 1–13.

Jain, P., Coogan, S., Subramanian, S., Crowley, M., et al. (2020). A review of ML applications in wildfire science and management. *Environmental Reviews, 28*(4), 478–505.

Kingma, D., & Welling, M. (2019). *An introduction to variational autoencoders*. Now Publishing.

Kovner, A. (2022, September 15). *Earthquake safety, one shake simulation at a time*. Berkeley Lab.

Lefohn, A. (2023, May 2). Latest NVIDIA graphics research advances generative AI's next frontier. *NVIDIA Blogs*.

Mandel, J., Amram, S., Beezley, J., Kelman, G., et al. (2014). Recent advances and applications of WRF-SFIRE. *Natural Hazards and Earth System Sciences, 14*(10), 2829–2845.

Mell, W., Jenkins, M., Gould, J., & Cheney, P. (2007). A physics-based approach to modelling grassland fires. *International Journal of Wildland Fire, 16*(1), 1–22.

Mildenhall, B., Srinivasan, P., Tancik, M., Barron, J., et al. (2020). NeRF: Representing scenes as neural radiance fields for view synthesis. In A. Vedaldi, H. Bischof, & J.-M. Frahm (Eds.), *European conference on computer vision* (pp. 405–421). Springer.

Niu, C., Phaneuf, M., & Mojabi, P. (2023). A diffusion model for multi-layered metasurface unit cell synthesis. *IEEE Open Journal of Antennas and Propagation, 4*, 654–666.

Parkes, B., & Demeritt, D. (2016). Defining the hundred-year flood: A Bayesian approach for using historic data to reduce uncertainty in flood frequency estimates. *Journal of Hydrology, 540*, 1189–1208.

Radke, D., Hessler, A., & Ellsworth, D. (2019). FireCast: Leveraging deep learning to predict wildfire spread. In S. Kraus (Ed.), *IJCAI* (pp. 4575–4591). IJCAI.

Schmidt, V., Luccioni, A., Teng, M., Zhang, T., et al. (2022). ClimanteGAN: Raising climate change awareness by generating images of floods. In K. Hofman & A. Rush (Eds.), *International conference on learning representations* (pp. 1–27). ICLR.

Scott, J., & Burgan, R. (2005). *Standard fire behavior fuel models: A comprehensive set for use with Rothermel's surface fire spread model.* Technical report, US Department of Agriculture & Forest Service.

Sharples, J., & Hilton, J. (2020). Modeling vorticity-driven wildfire behaviour using near-field techniques. *Frontiers in Mechanical Engineering, 5*, 69.

Storey, M., Bedward, M., Price, O., Bradstock, R., & Sharples, J. (2021). Derivation of a Bayesian fire spread model using large-scale wildfire observations. *Environmental Modelling & Software, 144*, 105127.

Teng, J., Jakeman, A., Vaze, J., Croke, B., et al. (2017). Flood inundation modelling: A review of methods, recent advances and uncertainty analysis. *Environmental Modelling & Software, 90*, 201–216.

Tirado Cortes, C., Thurow, S., Ong, A., Sharples, J. J., et al. (2023). *Analysis of wildfire visualisation systems for research and training* (pp. 1–20). *IEEE Transactions on Visualization & Computer Graphics.*

Wahlqvist, J., Ronchi, E., Gwynne, S., Kinateder, M., et al. (2021). The simulation of wildland-urban interface fire evacuation: The WUI-NITY platform. *Safety Science, 136*, 105145.

Wang, Z., She, Q., & Ward, T. (2021). Generative adversarial networks in computer vision: A survey and taxonomy. *ACM Computing Surveys, 54*(2), 37.

Weber, R. (1991). Modelling fire spread through fuel beds. *Progress in Energy and Combustion Science, 17*(1), 67–82.

Xiong, W., Luo, W., Ma, L., Liu, W., & Luo, J. (2018). Learning to generate time-lapse videos using multi-stage dynamic generative adversarial networks. *Computer Vision & Pattern Recognition. Arxiv.org*. https://t1p.de/7nr9x. Accessed 17 Dec 2023.

Yang, S., Lupascu, M., & Meel, K. (2021). Predicting forest fire using remote sensing data and ML. In *AAAI conference on AI* (pp. 14983–14990). AAAI.

Zhang, J., Xu, C., Liu, L., Wang, M., et al. (2020). DTVNet: Dynamic time-lapse video generation via single still image. In A. Vedaldi, H. Bischof, T. Brox, & J.-M. Frahm (Eds.), *European conference on computer vision* (pp. 300–315). Springer.

Chapter 5
Simulation of Extreme Fire Event Scenarios Using Fully Physical Models and Visualisation Systems

Khalid Moinuddin, Carlos Tirado Cortes, Ahmad Hassan, Gilbert Accary, and Frank Wu

Abstract Although extreme wildland fires used to be rare events, their frequency has been increasing, and they are now causing enormous destruction. Therefore, understanding such fire events is crucial for global ecological and human communities. Predicting extreme fire events is an imperative yet challenging task. As these destructive events cannot be investigated via experimental field studies, physical modelling can be an alternative. This chapter explores the capability of fully physical fire models to simulate these events and the potential of integrating these simulations with advanced visualisation systems supported by machine learning. By presenting case studies and future directions, we emphasise the potential and necessity of this integration for improved fire management and policy making.

Keywords Canyon fire · Computational fluid dynamics · Data visualisation · Extreme fire · Junction fire · Large-eddy simulation · Physics-based model · Wildfire modelling · Wildfire visualisation

---

K. Moinuddin (✉) · A. Hassan
Institute for Sustainable Industries & Liveable Cities, Victoria University (VU), Melbourne, VIC, Australia
e-mail: khalid.moinuddin@vu.edu.au; ahmad.hassan6@live.vu.edu.au

C. Tirado Cortes · F. Wu
iCinema Research Centre, The University of New South Wales (UNSW), Kensington, NSW, Australia
e-mail: c.tirado@unsw.edu.au; frank.wu1@unsw.edu.au

G. Accary
School of Engineering, Lebanese American University, Byblos, Lebanon
e-mail: gilbert.accary@lau.edu.lb

D. Del Favero et al. (eds.), *Climate Disaster Preparedness*, Arts, Research, Innovation and Society, https://doi.org/10.1007/978-3-031-56114-6_5

5.1 Introduction

Wildland fires, fuelled by climate change and human activities, have increased in frequency and intensity. Most of these fires are limited in intensity and do not cause significant harm. On the other hand, extreme wildfire events can have catastrophic impacts on ecosystems, infrastructure and human lives. Therefore, it is important to understand these extreme events, which are dynamic and propagate with high speed and intensity. This includes understanding why such fires occur and the risks they pose, creating situational awareness and developing preparatory strategies.

It is almost impossible to develop such understanding via experimental studies—primarily because they are too dangerous and may lead to actual fire events that imperil the safety of researchers. Moreover, the equipment and probes to measure various parameters can be monetarily expensive and may get damaged during experiments. Additionally, many important parameters such as fire intensity, flame length, angle and height, flow direction and turbulence are difficult to measure. Fully physical fire models, though computationally expensive, have emerged as viable tools for studying such extreme events. These models include turbulent fluid motion, all modes of heat transfer (conduction, convection, radiation), pyrolysis (gasification of fuel from the solid state before taking part in combustion), combustion, soot production and firebrand transportation. All required physics and chemistry are accounted for when simulating fire-fuel and fire-atmosphere interactions. As a result, physical modelling offers insights into the intricate dynamics of extreme fire events and accurately predicts risk. Yet, the complexity of these models requires intuitive visualisation systems to make insights accessible and actionable and interpretations meaningful.

However, a system capable of rendering these complex datasets requires a sophistication not found in traditional weather visualisation systems. On the data processing side, the sophistication in size and dimensions requires extra handling steps. On the data presentation side, there is the need to represent the data in more than two dimensions (2D). Current representation standards struggle to capture a crucial characteristic of the fire phenomenon: their durational dynamics.

5.2 Extreme Event Scenarios

Extreme fire events are characterised by their unpredictability and severe impacts, having historically caused significant losses, especially in urban settings or in the form of fire tornadoes. These events challenge our predictive capabilities and response mechanisms. Historically, extreme fires were rare. However, in recent years, their frequency has been increasing and causing considerable social, economic and environmental catastrophes.

Extended fire seasons are anticipated to result in extreme fires becoming a regular occurrence (Di Virgilio et al., 2019). In 2003, two large bushfires (the McIntyre's

Hut fire and the Bendora dam fire) merged and advanced on Australia's capital, Canberra, causing severe damage (Sharples et al., 2012). In 2017, several active fire lines in central Portugal merged into a wildfire. They initiated a violent firestorm that cost 66 people their lives (Pinto et al., 2022). The 2018 wildfire season in the US state of California set an unprecedented record, with 95 fatalities and the destruction of over 22,000 structures. Its devastation is largely attributed to dynamic fire behaviours such as merging (Filkov et al., 2019). During Australia's 2019–2020 Black Summer, the fire near New South Wales's Badja Forest Road in Countegany merged with multiple other fires—including the Big Belimbla Creek and the Dampier State Forest fire (turning first into the Big Belimbla Creek fire) and then merging with the Bumbo Creek fire. This created the momentous Badja Fire Complex, which affected various regions of the NSW South Coast. Fire merging was a critical factor in catalysing destructive winds and, in one case, led to the formation of a fire-generated vortex (Peace et al., 2021).

Predicting and preparing for such events presents numerous challenges. One of the primary difficulties is the limited historical data that could provide a robust foundation for a fast and mostly accurate prediction model. The impacts of global climate change further complicate matters as they alter traditional fire patterns and intensify fire risks, increasingly making empirical prediction models obsolete. The intricate interactions between factors such as vegetation, weather, topography and human activities add complexity to fire predictions, which require either full-physical modelling or parameterisation of physical modelling into empirical prediction models. Moreover, as urban areas increasingly encroach on wildlands, the potential for devastating fires affecting human populations increases, requiring intricate fire management and evacuation planning. Coordinating responses across multiple jurisdictions, each with its unique set of protocols and priorities, requires detailed mapping and collaboration, ensuring that communities are adequately protected from the unpredictable nature of wildfires (Davis et al., 2021).

Resource constraints pose another significant challenge. Maintaining readiness for extreme events, especially when frequent smaller fires already stretch resources, is daunting. Furthermore, conveying the risk of these uncommon but extreme events to the public is often challenging due to comprehension of low probability events, variable risk perception and desensitisation to warnings—to name but a few factors. This leads to potential complacency and a lack of preparation (Mackie, 2014; Hanson-Easey et al., 2019). Raising public awareness through powerful visualisation can be a key solution.

5.3 Physical Fire Models: An Overview

Forest fire models have been developed since the 1940s and differ widely in complexity. They can be divided into three categories: empirical, semi-empirical and physics-based.

Empirical models use past experiences and intuition to predict the behaviour of future fires. Semi-empirical models evaluate the properties of a steady "surface fire through a homogeneous solid-fuel layer, [such as the rate of spread (RoS) and the flames' height], based on an energy balance written in an inertial reference frame [attached to] the fire front (Rothermel, 1972). The main advantage [of such models] is [their] simplicity" (Morvan et al., 2022), which is why the Rothermel model is applied in FARSITE (Finney, 1998), the world's most widely used operational tool. FARSITE is a 2D fire propagation model that deploys a vegetation layer that accounts for a terrain's topography. It includes a vegetation library covering various ecosystems (e.g. grass, litter, shrubs, etc.). However, it is important to note "that the experiments used to calibrate the constants [of semi-empirical models] were performed only at a small scale, [i.e. in] solid fuel litters. For various reasons ([including] compactness of the fuel layer, low level of turbulence, [dimensions of the] vegetation layer, etc.), the conditions [of] such experiments [prevent the use of semi-empirical models for all configurations observed at field scale. This] motivated different research [teams to] couple a simplified fire-spread model (such as Rothermel's) with a [meso-scale atmospheric model (e.g. Filippi et al., 2011). This approach is considered very promising] for operational applications [that require] the simulation of wildfire propagation [at a] regional scale. [It is still being refined with] new (more physical) fire propagation models" (Morvan et al., 2022) emerging, such as elaborations of Balbi et al.'s model (2009).

More fundamental research is needed to foster a deeper and more nuanced understanding of wildfire behaviour and its underlying physics. Within limits, this may be developed via experimental fires. Yet, numerical simulation using fully physics-based models shows more promise. Examples of these are FIRETEC (Linn & Cunningham, 2005), FireStar3D (Morvan et al., 2018), FireFOAM (Edalati-nejad et al., 2022) and FDS (McGrattan et al., 2023). "Most of these models are based on a [multi-phase] formulation, assimilating the vegetation [cover] to a sparse, porous [medium] and [solving] a set of [balance] equations governing the behaviour of the coupled system formed by the vegetation and the surrounding ambient air" (Morvan et al., 2009). Grishin (1997) initiated this approach.

Physics-based models consist of two sets of differential equations, each describing the evolutions of the fluid and the solid mediums, coupled through interaction terms in the mass, momentum and energy equations. The first set of differential equations describes the evolution of the composition of the vegetation (i.e. solid fuel) as it is "subjected to the intense heat flux coming from the flaming zone" (Hassan, 2022). The second set of partial differential equations describes the evolution of a turbulent-reactive fluid flow, which results "from the mixture of the pyrolysis and combustion products with the ambient [atmospheric air]" (Morvan et al., 2018). Due to this high level of complexity, physics-based approaches are currently "limited to simulating fire behaviour at [local field] scale (i.e. a few hectometres). [They are only applied to studies that aim to improve] knowledge of wildfire dynamics (Frangieh et al., 2020) and to fire safety engineering studies [of structures located] at the wildland-urban interface. Their degree of complexity also increases the level of uncertainty of the [predicted results derived from physics-based

Table 5.1 Summary of main characteristics of four fully physical fire models

	FireStar3D	FDS	FIRETEC	FireFOAM
Solver	3D implicit	3D explicit	3D explicit	3D implicit
Low Mach model	Yes	Yes	No	Yes
Turbulence	TRANS/LES	LES	LES	LES
Turb./rad. interaction	Yes	Yes[a]	Yes[a]	No
Combustion model	Yes	Yes	No[b]	Yes
Multiple-fuel model	Yes	Yes	No	Yes
Small-scale	Yes	Yes	No	Yes
Field-scale	Yes	Yes	Yes	Yes

[a]The radiation heat transfer is increased empirically
[b]Pyrolysis and combustion occur at the same location without transport into the gaseous phase (as discussed in Hassan, 2022)

models]. Therefore, as in other [computational fluid dynamics] applications, it is [necessary] to enforce the confidence [attribute to the] results obtained" (Morvan et al., 2022) by regularly comparing them to experimental data (e.g. Frangieh et al., 2018; Hassan et al., 2023). An overview of the differences and similarities between fully physical fire models used in the literature is summarised in Table 5.1 (Morvan, 2011).

5.4 Visualisation Techniques Used in Physical Modelling

Given the complexities of physical models for different weather phenomena, there is a real need to push the limits of scientific visualisation to meet user needs. For example, the modelling software *FireStar* (Morvan & Dupuy, 2001), a simplified version of *FireStar3D*, generates 2.5D plots by animating traditional 2D plots. These representations allow the visualisation of changes in flow fields, such as velocity or temperature. This configuration provides some improvement over traditional 2D visualisations. However, they are limited when representing data derived from mathematical functions (Tirado Cortes et al., 2023).

Similarly, FireStar3D and FDS allow 3D visualisation of weather phenomena such as heat release, smoke and particles. This representation enables a multi-perspective view, which provides a superior analysis of weather phenomena that 2D representations cannot provide (Kraus et al., 2020). These tools also animate the outputs to increase the information communicated to the user. Still, new issues arise, such as interaction design and how it can significantly improve the user experience.

In both instances, immersive visualisation provides superior capabilities. First, it places the user inside the event, allowing them to interact with the environment through a multisensory experience (Ens et al., 2021). Second, viscerally experiencing a weather event such as a wildfire permits users to explore data meaningfully and to develop insight into the mathematical formulas that determine these events

(Lee et al., 2021). Immersive visualisation pushes for a realistic representation of the data, allowing observation of fire behaviours that easily get lost when data is abstracted and condensed for non-immersive setups (Marriott et al., 2018).

5.5 Case Studies

We have explored several specific modelling and visualisation scenarios. Their dynamic modes of fire propagation are shown in Fig. 5.1. We simulated four configurations of merging fires using two fully physical models: FireStar3D and FDS. The fire line lengths of the V arms are 50 metre. The scenarios include merging two-line fires, fire coalescence and eruptive fire behaviour in canyons. We used FireStar3D to model a surface fire and its propagation in Erica shrubs litter. On the other hand, we used FDS to model fire propagation in raised fuel (involving 6-metre-high Douglas fir trees with 4-metre crowns, standing on 0.5-metre-high Kerosene grass). The fuel types are characterised by specifying their thermo-physical and flammability properties.

RoS and energy (i.e. heat) release are the two important factors determining a fire's degree of devastation. Fast fire propagation and heat intensity can quickly engulf houses and communities at a moment's notice. As a result, built structures

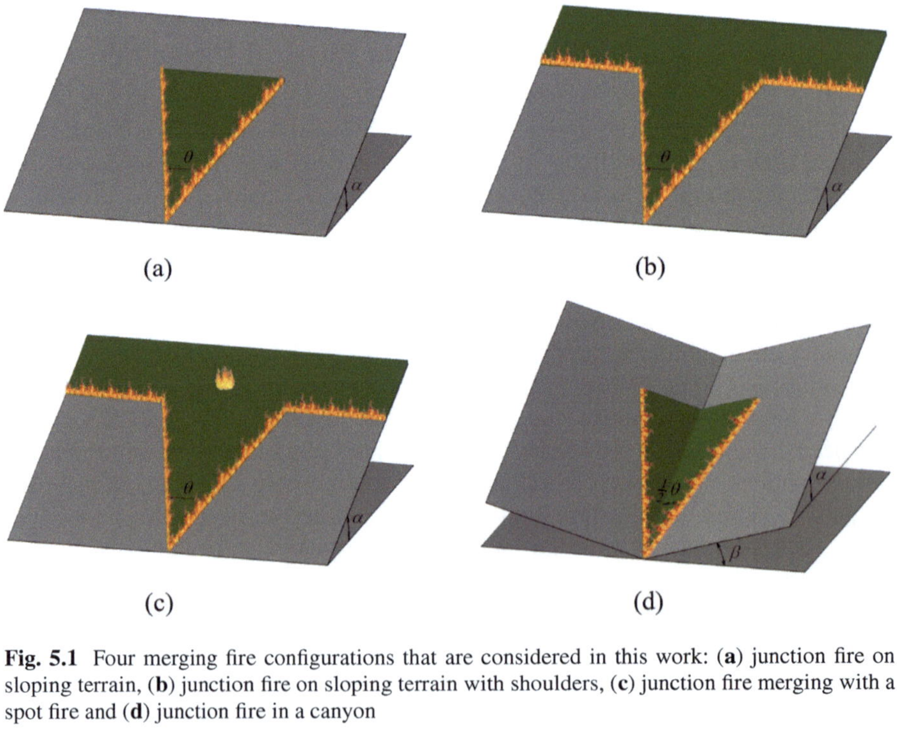

Fig. 5.1 Four merging fire configurations that are considered in this work: (**a**) junction fire on sloping terrain, (**b**) junction fire on sloping terrain with shoulders, (**c**) junction fire merging with a spot fire and (**d**) junction fire in a canyon

and human and animal lives can be lost. Smoke production can hinder evacuation efforts due to low visibility and cause long-term health effects due to its release of toxic particles. These three aspects are presented quantitatively and/or qualitatively in this chapter.

Figure 5.2 presents junction point propagation (left axis) and time evolution of the total heat release rate (HRR—right axis) obtained using the FireStar3D model. Using Tecplot software (2023), fire spread is visualised in Fig. 5.3 using the distribution of the solid-fuel bulk density at the vegetation cover surface. Flames are visualised by an iso-value "surface of the soot volume fraction [coloured] by the gas temperature. An [iso-value] surface of water-[vapour] mass-fraction [visualises smoke]" (Badlan et al., 2021). We can observe that up to ~15 s, the junction point propagates identically for cases (a–c). After that, case (a) with no shoulder propagates faster than the "with-shoulder" scenario (i.e. case b). We can observe a sudden jump in case (c), as the spot fire merges with the junction fire. The initial peak in the HRR is due to the ignition phase. As cases (b) and (c) had the same amount of burnable vegetative fuel, almost identical HRR is observed, indicating that the spot fire's contribution is not significant. The canyon fire (i.e. case d) propagates much slower

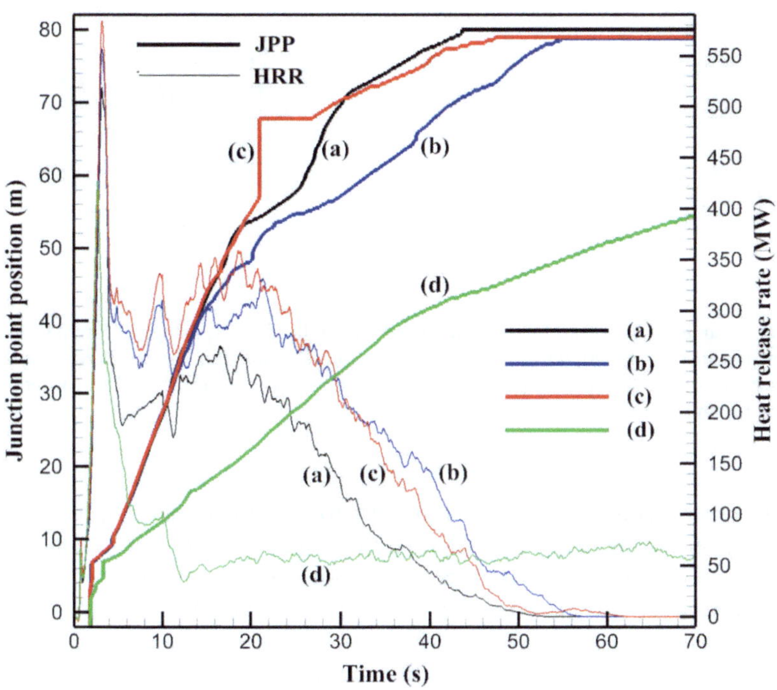

Fig. 5.2 Time evolution of the junction point position (JPP) and of the total HRR obtained in the four configurations shown in Fig. 5.1, using the FireStar3D model: (**a**) junction fire on sloping terrain, (**b**) junction fire on sloping terrain with shoulders, (**c**) junction fire merging with a spot fire and (**d**) junction fire in a canyon

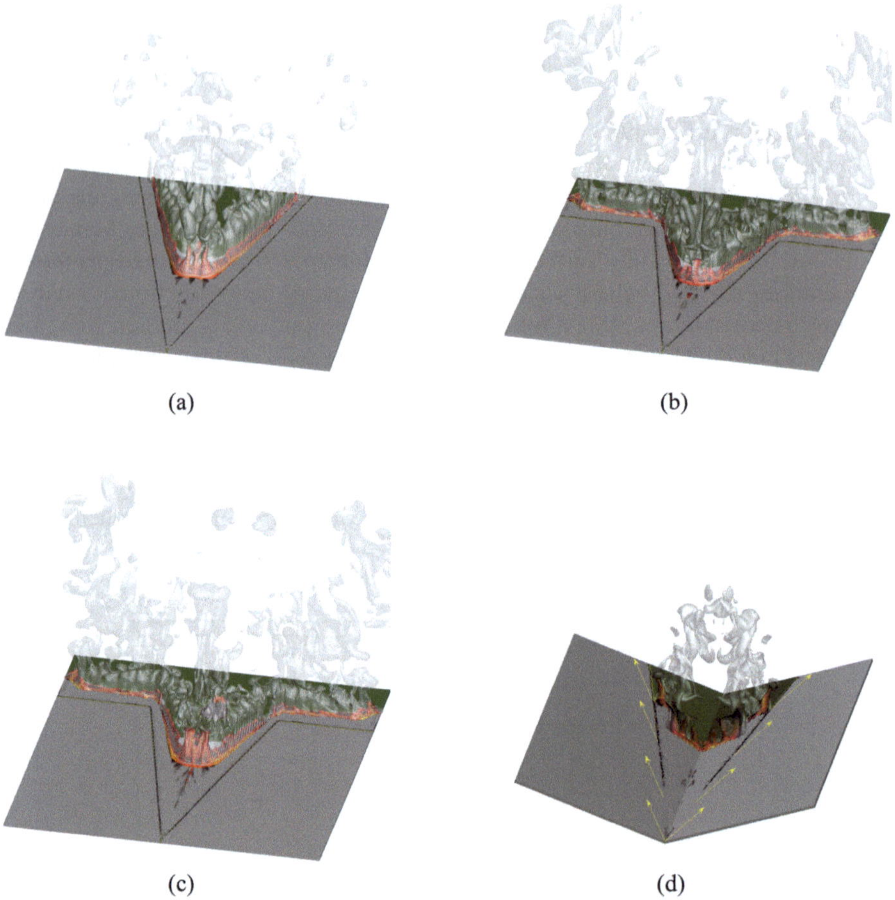

Fig. 5.3 Visualisation of four fire configurations shown in Fig. 5.1, using the FireStar3D model: (**a**) junction fire on sloping terrain, 15 s after ignition; (**b**) junction fire on sloping terrain with shoulders, 15 s after ignition; (**c**) junction fire merging with a spot fire, 15 s after ignition; and (**d**) junction fire in a canyon, 60 s after ignition

than the others. This is because a canyon fire follows the lines of the steepest slope (shown by the arrows in Fig. 5.3d) intersecting with the junction arms, which blocks fire propagation in this favoured direction. Comparing (a) and (c), after 5 s, a significant reduction of HRR is observed for the canyon fire. We can also observe in Fig. 5.2 that flame intensity and heat release are much lower than in the other cases.

With the FDS model, the canyon fire scenario was not modelled because the model currently does not have a clear ability to simulate slopes in multiple directions. The other three scenarios show similar quantitative behaviour (regarding JPP and HRR) observed in surface fire scenarios simulated by FireStar3D.

Figure 5.4 shows scenarios after 25 s from ignition that are visualised using the FDS companion software Smokeview. It shows junction fires represented by ignition lines. Flames and smoke are visualised by iso-value surfaces of the heat release rate per unit volume (>80 kW/m^3) and smoke extinction opacity (1000 m^2/kg), respectively. We can observe merging behaviour with intense flaming and large smoke billowing. If we set smoke extinction opacity to 8000 m^2/kg (i.e. the default value of Smokeview), more dramatic smoke billowing can be observed. However, it will obscure flame propagation.

Using Smokeview, animated 2D plots of flow fields (i.e. velocity, temperature, vectors, etc.), which are gas phase parameters, can be visualised. Similarly, we can visualise animated 2D plots of solid phase parameters, such as heat flux and temperature. These can provide insight into the physical phenomenon associated with fire propagation. In Fig. 5.5, 2D temperature plots along the domain's centre line are presented, giving viewers an idea of properties such as flame inclination, height, temperature distribution, etc. Additionally, 2D plots of radiation heat flux on the surface are presented, which can inform about the heating or cooling of the surface via radiation or convective processes. Legends for contours are presented in Fig. 5.5c. Figure 5.5 also includes 3D flames.

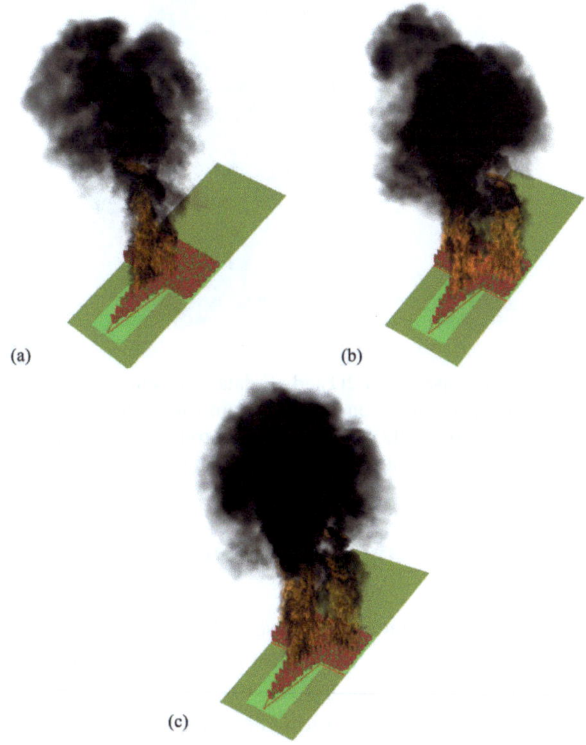

Fig. 5.4 Visualisation of the three configurations simulated using the FDS model: (**a**) junction fire, (b) junction fire with shoulders and (**c**) junction fire merging with a spot fire. All are on sloping terrain. Bright green and moss green surfaces represent burnable and non-burnable grass, respectively. Red and green particles represent tree crowns

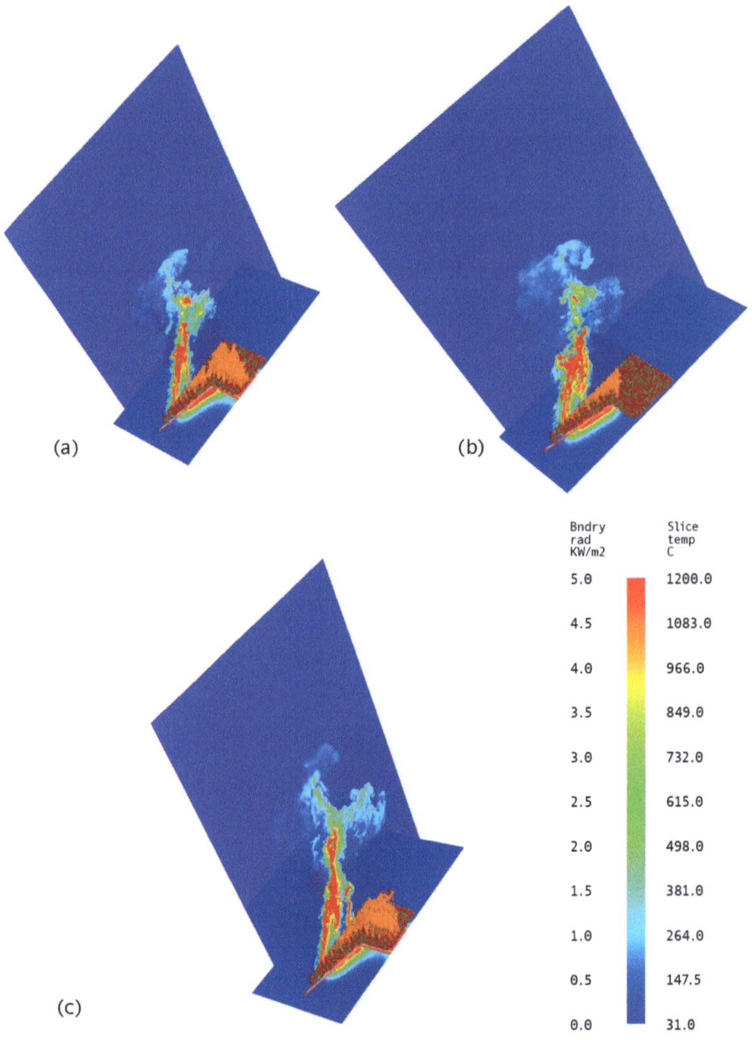

Fig. 5.5 Visualisation of 2D and 3D data from simulations, using FDS: (**a**) junction fire on sloping terrain, (**b**) junction fire on sloping terrain with shoulders and (**c**) junction fire merging with a spot fire. 3D flame data, 2D gas temperature through a vertical plane and 2D radiation heat flux data on the surface are presented. Time = 20 s from ignition

5.6 Integrating Simulations with a Visualisation System

A visualisation that can represent the complexities of such data is necessary to take full advantage of the recreation of wildfire events with physical models. However, the main challenge for visualising such models is the complexity of the data. For this reason, data integration methods are required to unlock new capabilities for powerful new visualisation approaches.

There are multiple benefits to improving the visualisation of physical models. First, they can facilitate new insight into the complex correlations between the natural environment and human ecologies. Second, they can illustrate scenarios that can educate and help plan damage control for at-risk communities. Finally, they inform society about the hypothetical future scenarios and how to prepare for them.

Physical models require a major processing complexity compared to other models (e.g. semi-empirical), given the high dimensionality of the data. Yet, these modelling systems are the only ones capable of plausibly reproducing wildfire behaviour (Badlan et al., 2021). Hence, an approach is needed to process their outputs into visually compelling form.

On the data side, the main limitation is the file size, which poses the biggest technological challenge when translating data on the manifold phenomena occurring during wildfires into visualisations. One potential solution is using Geostack (Hilton & Garg, 2022), a powerful geographic and weather data management tool. It is the core driver of SPARK (Miller et al., 2015), an empirical fire modelling application used widely across Australia. Geostack provides potent data handling and organisation utilities to manipulate the complex outputs of a physical model in ways that cater to visualisation. Software libraries such as Zarr, a Python programming language library, can support this translation by efficiently managing and compressing large data files.

On the visualisation end, *Unreal Engine* (UE) is the only tool capable of handling realistic representations of weather phenomena at such scale. UE is a game engine that has emerged as a new standard in VFX production. It excels at handling vast amounts of data without diminished performance. This commends it for application to physical modelling output. UE can recreate complex interactions and process multiple information layers without losing realism in its representation. Further, UE also facilitates the integration of multiple hardware, allowing a visualisation system to run across multiple platforms. This increases its appeal for a larger pool of potential users with different application needs, e.g. firefighters using an immersive-screen virtual-reality simulation trainer or fire behaviour analysts using the visualisation as part of their regular desktop-based reviews.

5.7 Future Directions

The current state of data visualisation and immersive technologies has led to the formation of immersive analytics as a new area of research (Dwyer et al., 2018). This area uses spatial interaction, collaboration and multisensory presentation to explore complex datasets, granting enhanced visualisation workflows compared to traditional 2D and 3D visualisation systems (Saffo et al., 2023). Future research should focus first on developing best practice conversion methods for the outputs of physical models so that an immersive analysis system can harvest them optimally. When designed right, such a system is poised to provide a superior learning experience compared to traditional workflows.

Another opportunity to explore in future research is enhancement of immersive visualisation using machine learning (ML). It promises to provide alternative approaches to overcoming the constraints of physical modelling, namely, the processing time and size of their data outputs. ML is already being used to improve fire modelling systems' accuracy and processing time by combining satellite images of active fire data with image processing-based predictions within geographic regions (Cheng et al., 2022). However, this image processing approach is limited because it must assume that all information and intricate physical processes represented by active fire satellite images are being encapsulated. This is not the case as it oversimplifies the complex physical processes and interactions involved in fire spread.

Another promising application domain of ML is the emerging modelling direction method known as physics-informed neural networks (PINNs). It involves the integration of fundamental physics principles into neural networks to enhance performance while reducing the demand for extensive data and computation time (Raissi et al., 2023). For instance, Ren et al. (2022) trained generative adversarial networks (Goodfellow et al., 2014) on large-eddy simulations. They achieved the incorporation of physical partial differential equation losses to simulate turbulent reactive flow. This could help speed up the processing times of physical models. Another application of PINNs in fluid dynamics involves an auto-encoder neural network model for interpolating low spatial-temporal resolution data to high resolution while adhering to physical constraints (Bode et al., 2021). This improves the efficiency of modelling of complex phenomena.

These ML methods attempt to reduce the barriers to applying physical modelling for visualisation. A line of research is already looking at how to apply these algorithms (Endert et al., 2017; Wang et al., 2022; Wang & Han, 2023). He et al. (2020) even presented a prototype pipeline from data collection and processing through ML training to visualisation. These all used simplistic and hypothetical simulations of fluids under very controlled scenarios. Future work should consider adapting these findings for wildfires and extreme weather event visualisation.

5.8 Conclusion

Extreme wildfire events are increasing in frequency, causing significant losses and challenging predictive capabilities and response mechanisms. Therefore, enhancing scientific insight and raising public awareness through wildfire visualisation is important. 3D physics-based models can provide the high-fidelity data required for such visualisation. For this chapter, we simulated four merging fire configurations using the FireStar3D and FDS models. The fires were visualised using obtained field variables, such as radiation heat flux, gas temperature fire intensity and smoke. While capable of reproducing wildfire behaviour, 3D physics-based models still pose challenges due to the complexity and size of their data. This also has repercussions for their visualisation. Tools such as Geostack hold promise since they have already been used for handling large amounts of data in other weather management applications, e.g. *SPARK*. Their full realisation depends on other cutting-edge technologies, such as UE, that can provide additional processing power. The VFX capabilities of the engine provide the necessary tools to recreate these datasets accurately. Finally, technologies such as ML can speed up the generation of new reliable data, overcoming the processing time limitations that currently afflict the work with physical models. Overall, using these methods at the data level of a visualisation system can support the creation of accurate representations while improving its overall performance. Eliminating the time-to-generate-data bottleneck can greatly improve the usefulness of the visualisation system and, thus, facilitate the generation of new knowledge. Finally, ML can be key to developing novel interaction methods, such as manipulating and reviewing data in real time during a visualisation session, which is currently not possible.

Acknowledgements This work was partially funded by the Australian Government through the Australian Research Council's Discovery scheme (FL200100004 and DP210102540) and was granted access to the HPC resources of Aix-Marseille Université.

References

Badlan, R. L., Sharples, J. J., Evans, J. P., & McRae, R. H. D. (2021). Factors influencing the development of violent pyroconvection. *International Journal of Wildland Fire, 30*(7), 484–497.

Balbi, J. H., Morandini, F., Silvani, X., Filippi, J. B., & Rinieri, F. (2009). A physical model for wildland fires. *Combustion and Flame, 156*(12), 2217–2230.

Bode, M., Gauding, M., Lian, Z., Denker, D., et al. (2021). Using physics-informed enhanced super-resolution generative adversarial networks for subfilter modeling in turbulent reactive flows. *Proceedings of Combustion Institute, 38*(2), 2617–2625.

Cheng, S., Jin, Y., Harrison, S., Quilodrán-Casas, S., et al. (2022). Parameter flexible wildfire prediction using ML techniques. *Remote Sensing (Basel), 14*(13), 3228.

Davis, E. J., Huber-Stearns, H., Cheng, A. S., & Jacobson, M. (2021). Transcending parallel play: boundary spanning for collective action in wildfire management. *Fire, 4*(3), 41.

Di Virgilio, G., Evans, J., Blake, S., Armstrong, M., et al. (2019). Climate change increases the potential for extreme wildfires. *Geophysical Research Letters, 46*(14), 8517–8526.

Dwyer, T., Marriott, K., Isenberg, T., et al. (2018). *Immersive analytics*. Springer.

Edalati-nejad, A., Sharples, J., Ghodrat, M. (Jan 2022). The effect of downslope terrain on wildfire dynamics in the presence of a cubic structure. In D. X. Viegas, & L. M. Ribeiro (Eds.), *Advances in forest fire research* (pp.775–783). IUC.

Endert, A., Ribarsky, W., Turkay, C., Wong, B. L., et al. (2017). The state of the art in integrating ML into visual analytics. *Computer Graphics Forum, 36*(8), 458–486.

Ens, B., Bach, B., Cordeil, M., Engelke, U., et al. (2021). Grand challenges in immersive analytics. In *Conference on human factors in computing systems – Proceedings*, p. 17.

Filippi, J. B., Bosseur, F., Pialat, X., Santoni, P. A., et al. (2011). Simulation of coupled fire/atmosphere interaction with the MesoNH-ForeFire models. *Journal of Combustion*, ID 540390.

Filkov, A. I., Duff, T. J., & Penman, T. D. (2019). Frequency of dynamic fire behaviours in Australian forest environments. *Fire, 3*(1), 1.

Finney, A. (1998). FARSITE: Fire area simulator, model development and evaluation. *USDA Forest Service.* Research paper INT-4.

Frangieh, N., Morvan, D., Meradji, S., Accary, G., & Bessonov, O. (2018). Numerical simulation of grassland fires behavior using an implicit physical multi-phase model. *Fire Safety Journal, 102*, 37–47.

Frangieh, N., Accary, G., Morvan, D., Meradji, S., & Bessonov, O. (2020). Wildfires front dynamics: 3D structures and intensity at small and large scales. *Combustion and Flame, 211*, 54–67.

Goodfellow, I., Pouget-Abadie, J., Mirza, M., Xu, B., et al. (2014). Generative adversarial networks. *Communications of the ACM, 63*(11), 139–144.

Grishin, A. M. (1997). *Mathematical modelling of forest fires and new methods of fighting them.* Tomsk State University.

Hanson-Easey, S., Hansen, A., Williams, S., & Bi, P. (2019). Communicating about heat-waves: risk perception, message fatigue, and threat normalisation. *PreventionWeb*. https://t1p.de/ek3sg. Accessed 29 Nov 2023.

Hassan, A. (2022). *Physics-based modelling of junction fires*. Master's thesis. VU.

Hassan, A., Accary, G., Sutherland, D., & Moinuddin, K. (2023). Physics-based modelling of junction fires: parametric study. *International Journal of Wildland Fire, 32*, 336–350.

He, W., Wang, J., Guo, H., Wang, K., et al. (2020). InSituNet: Deep image synthesis for parameter space exploration of ensemble simulations. *IEEE TVCG, 26*(1), 23–33.

Hilton, J., & Garg, N. (2022). Rapid geospatial processing for hazard and risk management using the Geostack framework. In *Proceedings of ISCRAM Asia Pacific conference*. ISCRAM.

Kraus, M., Angerbauer, K., Buchmüller, J., et al. (2020). Assessing 2D and 3D heatmaps for comparative analysis. In *Conference in human factors in computing systems – Proceedings*. ACM.

Lee, B., Brown, D., Lee, B., Hurter, C., et al. (2021). Data visceralization: Enabling deeper understanding of data using virtual reality. *IEEE TVCG, 27*(2), 1095–1105.

Linn, L., & Cunningham, P. (2005). Numerical simulations of grass fires using a coupled atmosphere – Fire model. *Journal of Geophysical Research, 110*, D13107.

Mackie, B. (2014). *Warning fatigue: Insights from the Australian bushfire context.* PhD thesis. University of Canterbury.

Marriott, K., Chen, J., Hlawatsch, M., Itoh, T., et al. (2018). Just 5 questions: toward a design framework for immersive analytics. In K. Marriott, F. Schreiber, T. Dwyer, et al. (Eds.), *Immersive analytics* (pp. 259–288). Springer.

McGrattan, K., McDermott, R., Vanella, M., Mueller, E., et al. (2023). *Fire dynamics simulator (Version 6.7.9), User's guide*. NIST.

Miller, C., Hilton, J., Sullivan, A., & Prakash, M. (2015). SPARK – A bushfire spread prediction tool. In R. Denzer, R. M. Argent, G. Schimak, & J. Hřebíček (Eds.), *IFIP advances in information and communication technology* (pp. 262–271). Springer.

Morvan, D. (2011). Physical phenomena and length scales governing the behavior of wildfires: A case for physical modelling. *Fire Technology, 47*, 437–460.

Morvan, D., & Dupuy, J. L. (2001). Modeling of fire spread through a forest fuel bed using a multiphase formulation. *Combustion and Flame, 127*, 1981–1994.

Morvan, D., Gavrilov, K., & Lyubimov, D. (2009). Physical phenomena governing the behaviour of wildfires: Numerical simulation of crown fires in boreal forest. In *19th Congrès Français de Mécanique,* Marseille, 24–28 August.

Morvan, D., Accary, G., Meradji, S., Frangieh, N., & Bessonov, O. (2018). A 3D physical model to study the behavior of vegetation fires at laboratory scale. *Fire Safety Journal, 101*, 39–52.

Morvan, D., Accary, G., Meradji, S., & Frangieh, N. (2022). 50 years of progress in wildland fire modelling. *Comptes Rendus. Mécanique*, online first, pp. 1–9.

Peace, M., Hanstrum, B., Greenslade, J., et al. (2021). *Coupled fire-atmosphere simulations of five Black Summer fires using the Access-Fire model.* BNH CRC.

Pinto, P., Silva, A., Viegas, D., Almeida, M., et al. (2022). Influence of convectively driven flows in the course of a large fire in Portugal: The case of Pedrógão Grande. *Atmosphere, 13*(3), 414.

Raissi, M., Perdikaris, P., & Karniadakis, G. E.. (2023). Physics informed deep learning (Part I). *Arxiv.org*. https://t1p.de/q243t. Accessed 17 Dec 2023.

Ren, P., Rao, C., Liu, Y., Ma, Z., et al. (2022). Physics-informed deep super-resolution for spatio-temporal data. *Journal of Computational Physics, 492*, 112438.

Rothermel R.C. 1972. A mathematical model for predicting fire spread in wildland fuels. *USDA Forest Service.* Research paper INT6115. Ogden: US Dept Agriculture

Saffo, D., Bartolomeo, S., Raynor, J., Di Bartolomeo, S., et al. (2023). *Unraveling the design space of immersive analytics: A systematic review* (pp. 1–12). *IEEE TVCG.*

Sharples, J. J., McRae, R. H., & Wilkes, S. R. (2012). Wind–terrain effects on the propagation of wildfires in rugged terrain: fire channelling. *International Journal of Wildland Fire, 21*(3), 282–296.

Tirado Cortes, C., Thurow, S., Ong, A., Sharples, J. J., et al. (2023). Analysis of wildfire visualisation systems for research and training. *IEEE TVCG*, 1–20.

Wang, C., & Han, J. (2023). DL4SciVis: A state-of-the-art survey on deep learning for scientific visualization. *IEEE TVCG, 29*(8), 3714–3733.

Wang, Q., Chen, Z., Wang, Y., & Qu, H. (2022). A survey on ML4VIS: Applying ML advances to data visualization. *IEEE TVCG, 28*(12), 5134–5153.

Chapter 6
Immersive Visualisation Systems as Alignment Strategies for Extreme Event Scenarios

Baylee Brits, Yang Song, and Carlos Tirado Cortes

Abstract Immersive systems are increasingly used to train first responders and prepare communities for extreme climate events. This chapter considers alignment issues that arise in their development and discusses how they might be resolved—taking as case study the *iFire* system currently being designed at the iCinema Research Centre. We particularly focus on ways to maximise the two non-moral values that define the success of any such system as well as of associated climate science: accuracy and verisimilitude. Drawing on the work of Shepherd et al. (2018) and Sharples et al. (2016), we theorise the epistemic and situational challenges to arrive at these values. Exploring solutions already proposed through the related concepts of 'storylines' (Shepherd, 2019), 'scenarios' (Lempert, 2013) and 'tales' (Hazeleger et al., 2015), we show how *iFire*'s values may be maximised through composition strategies derived from these concepts. Using this approach, we explain how *iFire* may 'simulate' links between uncertainty and affect to enhance decision making in uncertain circumstances. Our key findings are that alignment strategies for *iFire* are best described as 'interpretable' (rather than 'explainable') and can be achieved through qualitative methods. These describe compositional strategies deployed by the user that support reflective management of uncertainty.

Keywords Alignment · Immersive visualisation · Multi-agent systems · Scenarios · Storylines · Tales · Uncertainty

---

B. Brits (✉)
Office of National Intelligence, iCinema Research Centre, The University of New South Wales (UNSW), Kensington, NSW, Australia
e-mail: b.brits@unsw.edu.au

Y. Song
School of Computer Science & Engineering, The University of New South Wales (UNSW), Kensington, NSW, Australia
e-mail: yang.song1@unsw.edu.au

C. Tirado Cortes
iCinema Research Centre, The University of New South Wales (UNSW), Kensington, NSW, Australia
e-mail: c.tirado@unsw.edu.au

D. Del Favero et al. (eds.), *Climate Disaster Preparedness*, Arts, Research, Innovation and Society, https://doi.org/10.1007/978-3-031-56114-6_6

6.1 Introduction

This chapter considers alignment issues for immersive environments that stage extreme event scenarios for the purposes of preparation. To show what an aligned project might look like, we are using the *iFire* system, which is currently in development at the University of New South Wales's iCinema Research Centre. These sorts of environments have the potential to use visualisation and immersive experiences to inform perception, expectations and responses to wildfires (Hoang et al., 2010; Altintas et al., 2015). To create significant impact, these systems require a unique combination of accuracy as well as aesthetic and conceptual verisimilitude—a challenging combination in high-stakes ethical and decision-making situations. These systems are urgently needed because we are facing rapidly shifting fire landscapes around the world under a changing climate, which see a significant increase in extreme fire events. In response, AI capabilities are being developed to assist rate-of-spread (RoS) prediction and decision making in uncertain scenarios. The form of these systems and their potential usability for fire management requires an alignment model that ensures that the two key non-moral values—*accuracy* and *verisimilitude*—are maximised to achieve the overall goal for such systems: to generate insight and enhance preparedness for extreme event scenarios. This chapter provides a theoretical study of these alignment issues, developing an account of the efficacy of verisimilitude techniques, which range from 'storyline' to 'scenario' and 'simulation' theories—when they are placed in an interpretable framework. 'Alignment' is the guiding principle for circumscribing or shaping artificial intelligence (AI) actions to conform to human values, so that these systems successfully achieve human goals while minimising harm (Yudkowsky, n.d.). Alignment is a broad field that seeks to bring together intentions, values and contingencies. It is shaped by the nature of the impressions or assumptions that underlie our relations with AI—it is an "overarching research topic of how to develop sufficiently advanced machine intelligences, such that running them produces good outcomes in the real world" (Arbital, n.d.). Given the contestability of normative ethics, each alignment problem involves interrogating and defining what these 'good outcomes' might look like and the boundaries where the 'good' bleeds into the problematic, harmful and potentially catastrophic. The classic 'alignment story', whose origins are unclear, involves an AI machine that is designed to create paperclips, an apparently innocuous task. Through intelligence explosion and an optimisation process, this imagined paperclip machine creates a situation where it transforms "earth and then increasing portions of space into paperclip manufacturing facilities"—Armageddon in the name of wire squiggles (Yudkowsky, n.d.). This is a tale about the unknown capacities of an artificial general intelligence (AGI), and it is also, of course, a tale ad nauseam. Values are profoundly contingent, and value-oriented questions revolve around the modification, interpretation and application of multiple competing values and their individual limits. While this sort of imagined situation demonstrates the dangers of a naive application of goals in an exceptional AGI, it does not venture into the theatrically modest but conceptually more difficult territory of Janus-faced AI systems whose goals are also values: where a goal is inherently collaborative, as in multi-agent systems, or where the relevant values are not moral. This form of alignment can be achieved through an application of 'interpretable' strategies, which involve explicit articulation of qualitative techniques including, for example, 'insight'. Interpretability is a distinct problem of AI explainability and involves qualitative methods and case-by-case development. While AI explainability (XAI) is the process where AI decision-making processes are made transparent, interpretability refers to a broader, human-centred process that involves contextualised explanations. Interpretability recognises that any "ML decision is explained differently, depending on the person to whom it is explained, the reason why the explanation is needed, the place and time of the explanation, the ergonomics of human-machine interaction, and so on" (John-Matthews, 2022). The application of strategies theorised in scenario studies, psychology and climate science provides qualitative methods that can fulfil the need for interpretability and achieve alignment. This chapter surveys a series of methods, chief among these Theodore Shepherd's concept of 'storylines' (2019), which are defined as "physically self-consistent unfolding of past events, or of plausible future events or pathways" with no 'a priori probability' attached. In application to *iFire*, we couple this method with a theory of affect-informed 'simulations' of uncertainty (Anderson et al., 2019) and scenarios organised around perceived or discovered vulnerabilities (Lempert, 2013). Simultaneously, we show that an immersive environment has the capacity to offer the sort of "embedded experience" that climate science sorely lacks (Jasanoff, 2010; Shepherd & Lloyd, 2021).

6.2 Creating Accuracy and Verisimilitude for the Representation of Unpredictable Events

iFire assimilates geolocated data from fire simulation software to visualise wildfires in an immersive 360-degree cinema and other modules like 130-degree cinemas, single-projector displays, desktop computers and VR headsets. *iFire*'s goal is to accurately represent the nature of wildfires, using AI to predict and model rate of spread (RoS). It seeks to account for the dynamic and erratic characteristics of recent extreme fires, which can occur when "turbulent winds and mass spotting [...] create complex spread patterns [and when these interact and coalesce with] the main fire area" (Storey et al., 2021). This system will have the capacity to simulate changing fire behaviour based on shifting variables and user interactions. *iFire* is also developing an "AI framework that analyses, learns from and responds to individual and group behaviour in real time" in order to develop a multi-agent collaborative decision-making system that learns from past interactions (iCinema, n.d.). This system is intended to support preparedness efforts of a range of end users: scientists will be better able to analyse potential fire scenarios, firefighters can train and test strategies, and communities can increase awareness of vulnerabilities and improve preparedness strategies by experiencing extreme events in a safe environment. *iFire* attempts to model fires that are scientifically accurate and deliver verisimilitude (i.e. explanatory depth) by depicting a wildfire scenario that is convincing to both firefighters and scientists.

An AI 'misalignment' in this context, where the AI produces inaccurate or confusing visualisations, could lead to potentially dangerous real-world decisions or actions. Equally, an AI system that does not facilitate narrative plausibility for and elicits trust from its users, and which does not illuminate vulnerabilities or novel scenarios, will fail to create preparedness. Jasanoff (2010) has argued that scientific work on climate "arise[s] from impersonal observation" and, as such, can "detach knowledge from meaning". This is a unique alignment problem whereby these systems need to match accuracy with the sorts of verisimilitude that can provide "embedded experience" (ibid.). These problems are all the more pointed in the case of *iFire*, which aims to represent wildfires not only comprehensively but contingently, seeking to model a dynamic extreme firescape that is characterised by unpredictability. These dangers are not unique to the *iFire* system. It is comparable to Hädrich et al.'s mesoscale simulation of wildfires, a novel fire simulation system that replicates "dynamic behaviour and physics response of plant models" at forest scale using an innovative "module-based tree representation" (2021). This facilitates higher-fidelity simulation, because it can model realistic trees and variables, such as growth, as well as understanding feedback loops based on the heat radiating from burning trees. This project, however, is limited to a focus on trees and has no capacity to model grasses, undergrowth or leaf litter. Another comparable project is *VFire*, which is an immersive fire modelling system with similar aims to *iFire*. Although *VFire* is not AI enabled, it faces similar potential ethical conundrums as *iFire* because it does not include atmospheric circulation in its models and thus grapples with an inevitable partiality in the accuracy and realism of the fires (Hoang et al., 2010).

Users of the *iFire* system need to deal with significant novelty, including unexpected frequency, enormous scale and intensity of extreme fire behaviour, which can quickly escalate into hazardous and catastrophic scenarios, such as erratic firestorms (Sharples et al., 2016). This novel generation of wildfires (which are in many ways stoked by climate change) poses unprecedented risk and a complex set of challenges for the scientific community who grapples with their attendant uncertainty and novel experiential dimensions. The broad dilemmas of AI alignment here take on a particular pertinence for supporting decision making. The *iFire* system—and others like it—will be considered aligned if it is able to present users with an accurate and convincing fire simulation that offers them insight without being misleading. This involves maximising several competing aesthetic and scientific measures, which bridge fiction (i.e. the non-verifiable, non-confidence-based, intuitively plausible and meaningful scenarios created in immersive visualisation systems) and fact (the data-consistent, scientifically plausible, coherent, non-arbitrary scenarios) in a productive way to foster enhanced preparedness. In a situation characterised by uncertainty—both in climate science in general and in dynamically evolving extreme fire events in particular—, the *iFire* system needs to maximise the non-moral value of accuracy in a situation in which facts that are stable, retrospective and confidence-based might not exist.

6.3 The Challenge of Accuracy in Climate Science

Climate science is characterised by uncertainty. Significant work has identified the varieties of existing uncertainties and proposed methodologies for nevertheless enhancing knowledge and decision making under these conditions. This constitutes the 'interpretable' question in climate science. Shepherd et al. (2018) have developed the concept of 'storylines' to create reliability where uncertainty cannot be mitigated. They identify the incompatibility between increasing demands for clear, actionable climate information and the inherent uncertainty of key forms of climate data. They are responding to the fact that most public-facing climate reports rely on frequentist data to make probability statements about climate change. Frequentist statistics is generally held in opposition to Bayesian statistics and deals with the probability of data based on a null hypothesis, using the limit of the frequency of data as its probability. To demonstrate the limits of traditional probabilistic approaches to climate science, Shepherd and Lloyd (2021) detail studies of atmospheric circulation, which are "inherently regional, and involve dynamics (Newton's second law) as well as thermodynamics", which they contrast with confidence-based studies of thermodynamics. Shepherd (2019) has emphasised that such studies do not meet the three criteria typical of climate models, which require them to be accepted by climate theory, found in observations and contained in modelling. This concurs in important ways with wildfire modelling. Sharples et al. (2016) have argued that current fire prediction models fall short because they "are predicated on the assumption that the rate of spread of a wildfire burning in a quasi-equilibrium state can be uniquely determined by the local conditions of fuel, weather, and topography". They argue that these models, which rely on a 'quasi-steady assumption', do not work in scenarios where the fire is not adequately explained or represented by environmental conditions (ibid.). Such fires, which do not achieve the equilibriums assumed by RoS models, are known as 'dynamic fires'. A subset of these is called 'extreme bushfires', which "are associated with a higher level of energy, chaos, and nonlinearity" (Sharples et al., 2016). Storey and his colleagues (2021) argue that there is an imperative to raise awareness about these types of fires, which place significant and unprecedented demands on firefighters.

Shepherd (2019) sees uncertainty arising from different sources, namely, from future climate forcing, from climate system response to this and from the internally variable manifestation of a local climate at a given point in time. Further, uncertainty can result from human actions (i.e. scenario uncertainty), from limits in knowledge (epistemic uncertainty) or from random interfering elements (aleatoric uncertainty), whose probability may be partially deducted (ibid.). It is the latter two types of uncertainty which Shepherd argues must be held as distinct. Epistemic uncertainty is 'subjective', because it relates to what we know and do not know, whereas aleatoric uncertainty is 'objective', because it relates to events independent of our knowledge. Both latter types of uncertainty are relevant to *iFire* and the broad goal of preparation for extreme events: epistemic uncertainty involves, broadly, the change in weather systems and extreme weather events under climate change, where frequentist predictions with confidence levels attached become harder to make. Aleatoric uncertainty is relevant to such wildfires, as they are characterised by unprecedented levels of inherent dynamism.

Shepherd (2019) claims to cut the 'Gordian knot' of climate change uncertainty by shifting the question asked by climate scientists: he argues that "the societally relevant question is not "What will happen?" but rather "What is the impact of particular actions under an uncertain regional climate change?"" This is another way of saying that the climate discussion needs to move "from the "prediction [space]" to the "decision space"" (ibid.), without expecting the former to be a precursor for the latter. Shepherd argues that the situational, epistemic and aleatoric uncertainties that he describes should not preclude decisions and that they make "subjectivity inevitable" (ibid.). Where objectivity is not possible because of epistemic uncertainties, there is an ethical imperative to avoid the "illusion of objectivity", which can actually "reduce transparency" (ibid.). This, too, is a key imperative for *iFire*: accuracy in fire representation and behaviour must be achieved but must not be synonymous with objectivity. The simulations must offer contingent, possible future scenarios. This move from the probability space to the decision space is facilitated by the fact that epistemic uncertainty can be represented "through a discrete set of (multiple) storylines—physically self-consistent, plausible pathways, with no probability attached" (ibid.). Shepherd distinguishes storylines from scenarios through their remove from probability, so rather than asking what *will* happen, he asks "what would be the effect of particular interventions" (ibid.). The uncertainties that *iFire* must deal with are both situational (depending on how humans intervene or events unfold), epistemic (climate change increases unpredictable fire behaviour) and aleatory (inherent dynamics of extreme wildfires). This alignment issue of how to create accuracy in the face of uncertainty can be approached via 'interpretability', where contingency is an inherent part of the way a user understands the immersive system.

6.4 From Epistemic to Affective Uncertainty

Shepherd speaks about uncertainty as a relation to knowledge or events and in terms of the limits of frequentist probabilities that can dominate reports from the IPCC and similar authorities. But there is an affective dimension to uncertainty that is particularly relevant to the immersive experience of *iFire*. Anderson et al. (2019) "define uncertainty [as] a mental state, a subjective, cognitive experience of human beings rather than a feature of the objective, material world", which results from a conscious awareness of ignorance, i.e. lack of knowledge. There is surely scope for both definitions of uncertainty to exist—uncertainty as a quality of knowledge and uncertainty as a subjective experience—especially given that Anderson et al.

Table 6.1 The relationship between sources of uncertainty, types of uncertainty and affinities with mental experience

Climate science (Shepherd, 2019)	Types of uncertainty (Shepherd, 2019)—accuracy based	Inflection of mental state (Anderson et al., 2019)—verisimilitude based
Uncertainty 'in future climate forcing'	Scenario	Complexity
Uncertainty in 'the climate system response to that forcing'	Epistemic	Ambiguity
Uncertainty 'in the actual realisation of climate for a particular time window'	Aleatory	Probability/ambiguity
Decision-making strategy	*Storylines*	*Simulations*

distinguish a variety of experiences of uncertainty that conform, to a certain extent, to Shepherd's scientific discrepancies: probability, ambiguity and complexity. They note that probability, which they identify with risk, stems from the "randomness or indeterminacy of the future" (ibid.). Ambiguity, on the other hand, "arises from limitations in the reliability, credibility, or adequacy of probability", and complexity is yet again different, because it arises from difficulty in comprehension of information, rather than qualities of the information (ibid.). The below table outlines the affiliations between these different varieties of uncertainty. These are not understood as equivalences but as expressing related varieties of uncertainty in two different registers, each of which can be delineated based on the goals of *iFire*: accuracy and verisimilitude (Table 6.1).

These approaches address one aspect of the alignment question for AI systems that contribute to climate change storylines. By shifting the question to effect of actions—from the 'prediction space' to the 'decision space'—the prospect of 'good outcomes' (that is, outcomes that align with values of accuracy and which support the facilitation of preparedness and resilience) increases. Yet, users need to be made aware of the application and implication of 'storylines' in this context and their relation to concepts of uncertainty. This extends to both affective and cognitive approaches as well as experiences of uncertainty.

6.5 Storylining and Other Techniques: Tales, Simulations and Scenarios

The concept of storylines resonates with a process that Hazeleger et al. (2015) frame as *Tales of Future Weather*. The authors show that the limits of the conventional methodology applied in climate science, namely, 'MCDT'—model[ling] the entire "climate system, correct[ing] for biases, downscale[ing] to the scales of interest and finally translat[ing] into terms suitable for application"—cannot be adequately responsive to future weather (ibid.). They suggest a "complementary methodology" that arguably can "more fully explore the uncertainty of future climate for decision-makers today" (ibid.). Their 'tales' approach extends Shepherd's 'storylines' by seeking to reveal uncertainties. By explicitly shifting away from Shepherd's 'prediction space', 'tales' can allow for decision making but also clarify the present uncertainties. These uncertainties do not necessarily have to correlate exclusively with extremes in intensity but can also extend to increases in frequency (ibid.). A 'tales' scenario might reveal uncertainties in either domain.

For Anderson et al. (2019), uncertainty in itself has consequences, regardless of its origin: it "can lead to suboptimal decision making, negative affect, diminished well-being [sic], and psychopathology" and demands research and action to mitigate. This is an important and easily overlooked point. While Shepherd (2019) deals with the need to be responsive to uncertainty in terms of methodology, Anderson et al. (2019) remind us that uncertainty itself requires an active response to increase preparedness, resilience and wellbeing. Their suggestion for the mitigation of the negative effects of uncertainty aligns in certain respects with Shepherd's own method of *storylines*. Anderson et al. highlight the connection between uncertainty, simulation and affect, explaining that:

> [M]ental simulations might represent the critical mechanistic link between uncertainty and affective responses: uncertainty invites simulation of possible situations, and simulation, in turn, generates affective responses. For instance, if someone learns they *might* have cancer, they simulate what they think it would be like to have cancer (e.g., painful symptoms, treatment side-effects, hair loss, and death), which in turn generates negative affective responses. (Anderson et al., 2019)

Here the use of the term 'simulation' does not refer to particular media but to the imaginative process that rehearses possible outcomes from a situation. These simulations are proposed as a mediator between uncertainty and affect, with the implication that different imaginative processes can create different responses to uncertainty. This theory of simulation is a useful supplement to theories of 'storylines' and 'tales', because it focuses on the affective dimensions of these strategies, which neither Shepherd et al. nor Hazeleger et al. theorise. Anderson et al. (2019) point out that affect can change perceptions of likelihood and risk posed by extreme events. The latter are multidimensional scenarios requiring users to interact with information and content, with their responses additionally determined by immediately preceding emotional states and individual temperament on the identification of risk and attendant decision-making options (ibid.). Hazeleger et al. (2015)'s work also reflects this by structuring their 'tales' in ways that generate higher levels of concern: they relate information on extreme weather to likely everyday user experiences, which "was found to be a statistically significant determinant of higher levels of concern" (ibid.). This powerful affective dimension needs to be carefully considered in any use of 'storylines'. Simulations can affect perceptions of risk and uncertainty based on emotional states (prior and developing). This presents both possibilities and dangers for the alignment of a system like *iFire*. Effects like familiarity, optimism, capability, readiness or awareness—all affective varieties of preparedness—can contribute to a 'good outcome'. Instability, pessimism or fear would not.

Again, this demonstrates the necessity of differentiating 'interpretability' from 'explainability'. Shepherd's theory of 'storylines' demonstrates the need for forecast strategies that are responsive to inherent uncertainties, but the 'storylines' that he theorises are abstract and text based. If these storylines were transformed into immersive, visually rich simulations, they have the capacity both to offer much more substantial investigation of plausible scenarios and the effect of human actions in these scenarios. They would also address the affective dimensions of the uncertainties and risks of extreme events. Such rich simulations could capture the reality of uncertainty theorised by Shepherd but also address the affective dimensions raised by Anderson et al. They could achieve workable accuracy under uncertainty if users are aware of the simulations' narrative contingencies and if they are involved in specifying its compositional priorities.

6.6 Storylines and Interpretability

Forecast strategies that deploy the storylines approach are closely connected to AI concepts of 'explainability' and 'interpretability'. In the field of AI ethics, these two terms have often been conflated. However, recent work endeavours to separate them in order to distil requirements for AI decision making (e.g. Marcinkevîcs & Vogt, 2023). Here, explainability is associated with answering questions such as 'Why did the AI make this specific prediction?' or 'What factors influenced the AI's decision?' Such questions target the Shepherdian 'prediction space'. Yet, if this system is to occupy a 'decision space', we need to ask interpretive questions, like 'What meaningful insights do we glean from this simulation?' or 'What does this reveal about vulnerabilities?' These are fundamentally qualitative questions that involve multi-domain information and knowledge that comes from an interaction between the simulation and users, such as scientists, firefighters or community members (Table 6.2).

One strategy to transition into the 'decision space' is to use vulnerability as a compositional priority. This has been theorised by Robert Lempert (2013), who works in an adjacent field to storylines and simulations. His definition of 'scenarios' deals with uncertainty and dynamism in that they are less confident about the future than probabilistic predictions (ibid.). Lempert taxonomises a variety of human factors that show whether a scenario is successful or not:

Table 6.2 Qualitative and quantitative applications of explainability and interpretability

Explainability	Interpretability
Prediction space (frequentist statistics)	Decision space (storylines, simulation)
Quantifiable risk	Qualitative vulnerabilities
Computing risk of different options	Preparation, concern

> i) the usefulness of information so that the intended users regard it as credible, legitimate, actionable, and salient; ii) the relationships among knowledge producers and users, helping these parties to engage in mutual learning and 'coproduction of knowledge' while increasing mutual understanding, respect, and trust; and iii) the quality of the decision, which should include all five elements described above and be regarded by the parties as having been improved by the support received. (2013)

Lempert (2013) refines his methodology, though, through a particular focus on scenarios built around vulnerabilities. These scenarios aim to understand where a given policy might fail and subsequently how to find solutions or alternatives. He argues that it is vital to pay attention to the 'task' that the scenario is created to fulfil, contrasting a "decision structuring task that involves defining the scope of the problem, the goals, and the options under consideration" with 'a choice task', which deals with existing options (ibid.). Lempert contends that scenarios that illuminate the former will not necessarily do so for the latter. However, scenarios that are structured to illuminate vulnerabilities, he proposes, can fulfil both of these criteria and as such are stronger compositions with more opportunity for insight. A scenario that is structured around vulnerabilities in a proposed policy would enable users to understand where their plans may fail or what the vulnerabilities might be in their strategies. This is one appropriate way to ensure that storylines and simulations address the important affective dimensions that are central to modulating 'human' relations to uncertainty and increasing concern in the user. A storyline that makes use of Lempert's approach to scenarios, i.e. which is 'task oriented' and uses vulnerabilities as a compositional priority, will enhance preparedness and manage uncertainty simultaneously, because it prioritises the user's knowledge and expectations. For *iFire*, this may involve approaching the immersive environment with a task that is articulated prior to the experience and, in particular, making use of 'revealing vulnerabilities' as particularly pertinent task. This centrality of users' priorities mitigates the epistemic complexities of uncertainty in extreme events because they dictate verisimilitude rather than the measurable likelihood of a fire event.

6.7 Conclusion

iFire, and similar visualisation systems, intervenes in a field characterised by uncertainty and dynamism to create preparedness for the future. *iFire* aims to maximise truthful representation in an immersive environment to stage wildfires that will aid preparedness for scientists, firefighters and community members. This is a complex and ambitious undertaking, because such a project needs to create plausible futures that are necessarily subjective—they cannot be certain or confidence-based—and involve high levels of dynamism. *iFire* does not create probability-based forecasts but plausible storylines (in the sense developed by Shepherd and other climate scientists), which facilitate accuracy without the need for confidence. This also fulfils the requirement for verisimilitude, given that user intervention in storyline composition will facilitate explanatory depth. This allows *iFire* to intervene in the dynamic and ever-changing context of adaptation to climate change and to support preparedness not by previewing what will happen in the future but by allowing participants to survey vulnerabilities, stage events, experience contingencies and understand the consequences or impact of various actions. As such, *iFire* also represents a solution to the problems raised by Jasanoff regarding reports like the IPCC's, which "detach knowledge from meaning" by neglecting "embedded experience" (2010).

In this sense, *iFire* departs from other fire visualisation systems by shifting its priorities away from a confidence-enabled representation of a fire to an accurate projection of a plausible future fire scenario that is structured around user priorities and fully involves the situational uncertainty of human action and the epistemic and aleatoric uncertainties of climate science. As an immersive environment, *iFire* exceeds Shepherd's definition of 'storylines' because it functions as a simulation, thus presenting an affectively significant version of a storyline. This too presents an alignment issue in that the affective outcomes of the simulation need to be aligned with preparedness rather than panic, as well as awareness rather than confusion. An interpretable system that is focused on ensuring users focus on qualitative categories like insight and revelation (rather than seeking an explanation of how the AI makes decisions) has the capacity to forge links between affective states and types of knowledge.

The AI components of *iFire* can create optimal outcomes by an explicit task-oriented or purpose-oriented focus, the premier example being Lempert's positioning of vulnerability as a compositional priority. This explicit articulation of both purpose and compositional priorities is central to an 'interpretable' AI system. Simultaneously, this investigation shows that immersive systems such as *iFire* have a reciprocal alignment function. *iFire*, when conducted through compositional methods from scenario modelling, can produce storylines akin to those theorised by Shepherd et al. But it does so with the capacity to present an affectively rich storyline that can link feelings with knowledge. As such, *iFire* delineates the role creative arts can play in fostering climate change knowledge and preparedness, i.e. in shaping compositional structures and interpretable systems so that they give meaning to novel and unpredictable scenarios.

Acknowledgements Baylee Brits' postdoctoral fellowship is funded by the Australian Government's Office of National Intelligence (RG220479). Carlos Tirado Cortes' postdoctoral fellowship is funded by the Australian Research Council (FL200100004, directed by Laureate Prof. Dennis Del Favero), which develops the *iFire* research. Yang Song and Baylee Brits are involved in this project as expert collaborators.

References

Altintas, I., Block, J., De Callafon, R., Crawl, D., et al. (2015). Towards an integrated cyberinfrastructure for scalable data-driven monitoring, dynamic prediction and resilience of wildfires. *Procedia Computer Science, 51*(1), 1633–1642.

Anderson, E. C., Carleton, R. N., Diefenbach, M., & Han, P. K. J. (2019). The relationship between uncertainty and affect. *Hypothesis and Theory, 10*, 2504.

Arbital. (n.d.). AI alignment. *Arbital.com*. https://t1p.de/uaufm. Accessed 24 Oct 2023.

Hädrich, T., Banuti, D. T., Palubicki, W., Pirk, S., & Michels, D. L. (2021). Fire in paradise: Mesoscale simulation of wildfires. *ACM Transactions on Graphics, 40*(4 (July)), 1–15.

Hazeleger, W., van den Hurk, B., et al. (2015). Tales of future weather. *Nature Climate Change, 5*, 107–113.
Hoang, R., Sgambati, M., Brown, T., Coming, D., & Harris, F. (2010). VFire: Immersive wildfire simulation & visualization. *Computers & Graphics (Pergamon), 34*(6), 655–664.
iCinema Centre. (n.d.). iFire project overview. *icinema.unsw.edu.au*. https://t1p.de/rb03f. Accessed 30 Oct 2023.
Jasanoff, S. (2010). A new climate for society. *Theory, Culture and Society, 27*(2–3), 233–253.
John-Matthews, J.-M. (2022). Some critical and ethical perspectives on the empirical turn of AI interpretability. *Technological Forecasting and Social Change, 174*(January), 121209.
Lempert, R. (2013). Scenarios that illuminate vulnerabilities and robust responses. *Climactic Change, 117*, 627–646.
Marcinkevičs, R., & Vogt, J. E. (2023). Interpretable and explainable machine learning: A methods-centric overview with concrete examples. *WIREs Data Mining and Knowledge Discovery, 13*(3), e1493.
Sharples, J. J., Cary, G. J., Fox-Hughes, P., Mooney, S., et al. (2016). Natural hazards in Australia: extreme bushfire. *Climatic Change, 139*, 86.
Shepherd, T. (2019). Storyline approach to the construction of regional climate change information. *Proceedings of the Royal Society A, 475*, 20190013.
Shepherd, T., & Lloyd, E. (2021). Meaningful climate science. *Climactic Change, 169*, 17.
Shepherd, T., Boyd, E., Calel, R., Chapman, S. C., et al. (2018). Storylines: An alternative approach to representing uncertainty in physical aspects of climate change. *Climatic Change, 151*, 555.
Storey, M., Bedward, M., Price, O. F., Bradstock, R. A., & Sharples, J. J. (2021). Derivation of a Bayesian fire spread model using large-scale wildfire observations. *Environmental Modelling & Software, 144*, 1–2.
Yudkowsky, E. (n.d.). Squiggle maximizer. *Lesswrong.com*. https://t1p.de/ei5i0. Accessed 24 Oct 2023.

Part II
Narrating

Chapter 7
Moving Beyond Recovery and Reconstruction: Imagining Extreme Event Preparedness Through Performing Arts

Jane W. Davidson, Sarah Woodland, Helena Grehan, Simonne Pengelly, and Linda Hassall

Abstract This chapter begins by examining the importance of resilience in response to extreme weather events, before considering the role of performing arts projects in dealing with the aftermath of disaster. Including First Nations approaches, examples from diverse cultural settings and the powerful potential of digital technology, it reveals how performing arts endeavours afford aesthetic opportunities that can give voice to and make sense of crisis experiences that precipitate mental health and wellbeing challenges as devastating as the events themselves. Drawing on critical research findings including those from the 10-year Beyond Bushfires study in Australia, the authors demonstrate how empathy and social bonding can be fostered through artistic engagements to develop personal and community resilience and support creative recovery. Forewarning the accelerating and intensifying unpredictable character of disasters driven by climate change, the authors go on to underscore the need for significantly greater preparedness for future crisis events. Engaging

J. W. Davidson (✉)
Faculty of Fine Arts and Music, The University of Melbourne, Parkville, VIC, Australia
e-mail: j.davidson@unimelb.edu.au

S. Woodland
Victorian College of the Arts, The University of Melbourne, Southbank, VIC, Australia
e-mail: sarah.woodland@unimelb.edu.au

H. Grehan
Western Australian Academy of Performing Arts, Edith Cowan University, Mount Lawley, WA, Australia
e-mail: h.grehan@ecu.edu.au

S. Pengelly
School of Architecture and Built Environment, The University of Newcastle, Callaghan, NSW, Australia
e-mail: simonne.pengelly@uon.edu.au

L. Hassall
School of Humanities, Languages and Social Science, Griffith University, Brisbane, Australia
e-mail: l.hassall@griffith.edu.au

D. Del Favero et al. (eds.), *Climate Disaster Preparedness*, Arts, Research, Innovation and Society, https://doi.org/10.1007/978-3-031-56114-6_7

with a small body of existing work, the authors investigate the possibilities of 'performing preparedness' – embracing storytelling, embodied performance practices and digital technology as routes to developing agency, empowerment, understanding and strategies that can build capacity for dynamic readiness in rapidly evolving and unforeseen crises.

Keywords Creative recovery · First Nations approaches · Participatory performing arts · Resilience · Preparedness

7.1 Introduction

The *UN Global Assessment Report on Disaster Risk Reduction* (UN Office for Disaster Risk Reduction, 2022) noted that the global climate emergency is growing. Catastrophic weather events in the last four years provide confirmation. Australia's Black Summer (December 2019 to February 2020) saw devastating impacts, including loss of human life. On the eastern seaboard, settlements, native and commercial forest and farming lands were annihilated, and millions of animals and plants were destroyed (Australian Institute of Health and Welfare, 2020). While Australia has a long history of extreme weather events, film footage of wildfires and severe flooding in Northern Europe in 2022 was shocking. Ten wildfires occurred near London, and some 800 individual wildfire episodes left unprecedented scars on England's landscape (Horton, 2022). In 2023, Canada experienced 6551 wildfires, burning nearly 50 million acres between March and October, with 346 fires deemed out of control (Cecco, 2023). These and other extreme weather events, all directly associated with climate change (Igini, 2023), represent the escalating worldwide crisis.

Emergency services globally acknowledge not only that extreme weather events will increase in frequency and severity but that associated disasters will become more commonplace and more dangerous (Kemp et al., 2022). As societies become increasingly culturally complex in terms of their interface with the natural world, the "command and control" practices that have dominated disaster management will become ineffective (Mullins, 2021). In the face of unpredictable extremes in weather, it is imperative to build preparedness, coping and recovery solutions that engender foresight, resilience, resource replacement and post-traumatic growth. Considering this, it is crucial that toolkits for imagining and coping with unpredictable future weather events be prioritised. In this chapter, we explore ways in which performing arts have been used to represent disaster and manage its aftermath, and we explore the embodied thinking and action these artistic forms can provide in developing future preparedness for an escalating climate emergency.

7.2 Performing Resilience and Supporting Post-traumatic Growth

In disaster and emergency scenarios, resilience as a capacity to withstand or to recover quickly has tended to refer to the post-disaster period, processes of successful behavioural flexibility and the ability to adjust to both internal and external demands of extremely challenging experiences (López-Marrero & Tschakert, 2011). Kenneth Ginsburg has identified seven components or behaviours (the "7 Cs") that together promote personal psychological resilience: "competence, confidence, connection, character, contribution, coping and control" (American Psychological Association, 2023). Natural disasters typically present unexpected, unimaginable and devastatingly traumatic events that easily undermine or even obliterate the ability to display these behaviours and prevent individual recovery and healing. Performing arts have had a powerful role to play in promoting Ginsburg's 7 Cs and in the development of personal resilience.

At a systemic level, socio-ecological systems theory (SES; see Holling, 1973) usefully articulates the "intertwined systems of people and nature embedded in the biosphere" that are affected in natural disaster scenarios. At a meta level, the shock a natural disaster delivers is followed by possible outcomes: systems stasis, systems development or systems failure. Recovery and resilience in this context support continuation and development to avoid systems failure. Given the interrelated social, economic and ecological impacts, it seems that understanding how resilience can be strengthened at the systems' level is required in preparation for and not just after natural disasters.

Folke has been a key researcher in the SES field, applying "resilience thinking" to "deal with complex adaptive system dynamics and true uncertainty and how to learn to live with change and make use of it" (2016). Indeed, Folke et al. (2002) describe "resilience building" as flexible, diverse, responsive and evolving mechanisms. "Adaptation and transformation" become the outputs of resilience, allowing for response to the sudden shocks that produce dynamic change. Adaptation is contextualised as the ability of a system to sustain development along current pathways following disruption. Transformation describes the shift to alternative pathways or the creation of entirely new ones. Resilience provides SES with the capacity to transmute crisis into opportunity and depends on finding moments of opportunity to navigate crisis conditions (Folke, 2016). Kendra and Wachtendorf (2003) have referred to resilience building as "not merely as the application of scientific knowledge and techniques, but also as an art" (ibid.). For them, resilience is "the product of a kind of craft skill, or an artistic interpretation and response to singular, unexpected, anomalous events as opposed to a rationalised predetermined response to what is regular or expected" (ibid.). Therefore, building or exhibiting resilience requires "organisational craftsmanship…[and]…artistry" (ibid.). So, even at the systemic level, artistic thinking and practices have strong potential in generating both adaptive and transformative solutions to the crisis conditions of disaster.

Across human history, embedded in different cultural contexts, performing arts have been used to express, regulate, control and transform self, community and society through diverse practices (Davidson & Garrido, 2014). Interactions between co-performers and audiences take place to produce highly significant events delivering profound consequences. In some contexts, crucial cultural knowledge can be transmitted or discovered only in the time and space dedicated to performance (Davidson, 2011). In more recent times, performing arts have been used to address climate issues, often serving to support coping, adaptation and even transformation, thus providing evidence of their potential in developing resilience (Demos, 2016). Interventions using music, theatre, dance and integrated forms have been used in fire, flood and tornado recovery settings (Egan & Quigley, 2015; Peek et al., 2016). Music therapy's focus on moving an individual or group through a psychological process that has explicit thinking, behavioural and/or physical benefits has been successfully employed with a range of different people in disaster scenarios. In one example, "children who had survived a tornado in the Southeastern United States" used music making to describe and express emotions associated with lived experiences of this event. The musical work also enabled them to imagine and then realise a safe transition back to school after the event, building both personal and community resilience (Davis, 2010).

In another music therapy project with teenagers "after the Black Saturday fires in Victoria, Australia in 2009", therapist researchers reported that the musical tasks of improvisation, song writing and discussions surrounding the production of musical works enabled young survivors to "hear one another", bond, share experiences of loss and subsequently regain confidence (McFerran & Teggelove, 2011). Indeed, during and after the Black Saturday fires more generally, community members used arts practice skills to work in local crisis centres with disaster-affected families, facilitating processes in which they were able to listen, process feelings, express fear and hope, collaborate in artistic acts to mourn and recover and find respite and relaxation in a safe space. Rita Seethaler was one music leader whose incredible efforts in singing and band work, in particular the creation of a steel band, were recognised in Arts Victoria's 2011 Recovery Program report (Fisher & Talvé, 2011; see also Garrido et al., 2015). Activities have continued and been adapted for 15 years since that particular disaster, some moving into everyday social groups and others deliberately focused on the journey from disaster to transformative resilience in memorial rituals where community survivors come together to share their thoughts and feelings at fire sites, singing, dancing, speaking and sharing visual artworks to honour and remember those people, animals and places that were destroyed.

Practices such as these can be seen with each new emergency. Extensive floods in Lismore, New South Wales, in 2022 destroyed over 400 houses, displaced more than 31,000 people and disrupted 3000 businesses, with water, sewage and power utilities extensively damaged (Gilmore, 2022). Arts practices provided extensive community recovery intervention. Artist Claudie Frock endured great trauma after she, her partner and her dog were rescued by the State Emergency Service in a boat, having been in water up to their chests. Frock, as a long-established artist, performer

Fig. 7.1 Creative recovery session at Lismore. (Image provided by Claudie Frock)

and arts worker in Lismore, responded by using aesthetic opportunities, working to connect people. She and others engaged in creative recovery sessions for people to come together, make collages and have space to write, recite poems and listen to and make music. She noted: "Programs like this are so important at the moment when people have become displaced and isolated from each other as a result of this disaster" (qtd. in Stephens, 2022) (Fig. 7.1).

Applied theatre, theatre in education and drama therapy have also addressed a range of issues after disaster, exploring community vulnerability and working to find routes to recovery and resilience through imaginative exploration, questioning and solution-finding (Heras & Tàbara, 2014; O'Connor, 2013). Typically devised in close collaboration with the disaster-affected communities, this work draws on a range of theatre-based methods to contribute realistic problem solving through the prism of a fictional story.

A pilot study of youth theatre in Australia found that "child and youth [performance] addresses the climate crisis across four intersecting domains: from disaster preparedness, through first response, to disaster recovery, and climate activism […]. Across these domains, [performance[traverses artistic [practices] and everyday actions that contribute to children and young people's [ability to process trauma and eco-anxiety]" (Woodland et al., 2023) and cultivate a sense of critical hope, agency and community (ibid.). This accords with findings from other studies in child and youth drama and performance in response to the climate crisis (Egan & Quigley, 2015; Freebody & Finneran, 2021; Gallagher et al., 2022).

The projects mentioned above have all offered positive gains to those attempting to manage personal and community impacts of disaster scenarios. Harms et al. (2023) aggregated and analysed individual and community data from the Beyond Bushfires and more recent 10-year Beyond Bushfires studies (3–4 years and 10 years after the Black Saturday fires of 2009). This resulted in tracing 391 individuals affected in 2009 and exploring the relationships between their individual demographics, bushfire exposure and community-level variables within the framework of post-traumatic growth (PTG). This theory argues "that people who endure psychological struggle following adversity can often see positive personal growth afterwards" (Tedeschi & Moore, 2021). In PTG, "people develop new understandings of themselves, the world they live in, how to relate to other people, the kind of future they might have and a better understanding of how to live life" (Collier, 2016). The study revealed that PTG was more pronounced for women and was more likely for those who had been in medium- to high-affected fire areas, rather than low-affected areas. These results might reflect gendered expectations around emotional expression and self-regulation (Davidson & Garrido, 2014), though positive PTG effects were found to occur at individual levels for all affected.

Projects using performing arts present platforms to strengthen and build psychological resources like resilience and personal growth and empower people to manage the very severe adversities natural disasters present. The Creative Recovery Network (CRN) has been operating in Australia since 2011, providing resources, training and a community of practice for the use of arts-based programmes within disaster management recovery. The CRN recognises and supports First Nations approaches, and in their podcast series *Creative Responders*, a 2020 episode focuses on First Nations knowledges to understand the catastrophic impacts of rising water temperatures on plant and animal life in the Great Barrier Reef. Close to Gimuy Walubara Yidinji Country and on Gunggandji Sea Country, the First Nations interviewees discussed how they learned to respect and treasure the reef as they learned about its creation through cultural stories, involving dance and song. Dustin Maloney, a Kuku Yalanji and Yidinji man and Reef Tourism Guide, explained that in performing cultural stories for tourists, the message of a natural phenomenon to be respected and revered is strongly communicated, emphasising the importance of protecting the reef as a good place for providing food but also for the cultural flow or aesthetic contained within it, generating as he described: "A bit of a peace, peace inside of us".

In the same podcast, Aunty Gertie (Gertrude) Deeral, a First Nations artist of international acclaim from Hopevale Arts and Cultural Centre near Cairns in Northern Queensland, spoke of the processes of performance in cloth-making to bring community together physically and emotionally to "yarn", dye fabric, make clothes and be in the same emotional zone together. For Aunty Gertie, the performativity of this experience offers more than creative expression and social opportunity, and it is about finding memories, stories of the creation and relationship between people, place and nature, so important to generating and disseminating knowledge to make meaning and bring common purpose. While the conversation is about the protection of the reef, it could be applied to wildfires and flooding and their associations and deep histories in the Australian landscape that First Nations' cultural practices and beliefs support. Indeed, the concept of being "on Country" signifies much more than a simple physical presence, rather it offers a deep connection to the experience of land as a source of knowledge, culture and meaning and demonstrates a need to support and respect cultural practices, which promote an ongoing relationship with the land and its climate in ways most Australians have failed to recognise (Cameron, 2020).

7.3 Performing the Politics of Disaster Experience

In addition to performing arts interventions used specifically for recovery, work in professional theatre has taken a more politically focused awareness-raising approach. On 30 June 2010, the *Independent* newspaper in England reported on the increase in theatre works taking the "devastating power of nature as their theme". One such work was "*Katrina*, by Jonathan Holmes, which was staged over several floors in the Bargehouse in Oxo Tower Wharf", in London in 2009. The work focused on the impact of the 2005 floods in New Orleans, when Hurricane Katrina left 80% of the city submerged in water, resulting in 1836 human fatalities. The performance involved the audience moving from "a pre-hurricane tourist office into a bar that had been wrecked by flood water" (Billington, 2009).

The immersive performance incorporated "sounds of the hurricane and of the city of New Orleans falling apart", with director Holmes saying, "Sound is a great carrier of information...it can give a sense of what it might have been like to be there" (ibid.). Audience members were kept in a state of discomfort and – hearing verbatim testimony – were obliged to perch on stage "debris". Billington argued the work straddled political agitational and visceral experiential theatre, offering audiences the plight of survivors, examples of heroism and building a "staggering picture of official lies and ineptitude". This performance engaged empathy, raised awareness and perhaps even incited some affirmative action for the case of the survivors and the climate debate more generally.

Similar examples from professional theatre include *Life Streaming* commissioned by the London International Festival of Theatre (LIFT) and National Theatre London that enabled audience members to communicate via computers with the survivors of "the 2004 Indian Ocean earthquake and tsunami, one of the deadliest natural disasters in human history [with over a] quarter of a million [fatalities]" (Williams, 2010). Mark Ball, who commissioned the play, reported to Williams that this capacity to merge aesthetic engagement with factual information became very powerful "over the course of an hour you start to feel you can get to know the people behind those one-dimensional images of the disaster. You get a much more nuanced understanding of the impact that the tsunami had on a person's life." (ibid.)

As these examples show, there are many ways to interrogate our views of and responses to natural disasters that can help both survivors and spectators, challenge the agencies that in some cases have mishandled such events and potentially encourage climate action and activism.

7.4 Performing Preparedness

As illustrated above, climate-focused performing arts have focused on response, recovery and resilience. Surprisingly little has been done in preparing for these disasters, despite the ever-increasing emergency and the trend of growing eco-anxiety around extreme weather and its consequences (Kurth & Pinkala, 2022). Preparedness embraces the actions, measures and capacities built in advance to enhance the ability of individuals, communities and organisations to respond effectively to potential risks or challenges (Hémond & Robert, 2012). In the field of disaster management, it involves proactive planning, scenario modelling, resource allocation and disaster training to ensure readiness for emergencies (Madrigano et al., 2017). Given the power of the performative, it is surprising that preparedness is an under-explored area. Performing arts certainly offer symbolic forms of expression where people can begin to think about how we may face a much more extreme future regarding weather events (Thurow et al., 2023). They can experiment in imaginative and emotional ways to envision scenarios that may have been unimagined previously. This is critical, as preparedness is a function of the prior experience of a perceived threat (Lazo et al., 2015). The two professional theatre performances described above act in this way, highlighting the advantage of performing arts that individuals and communities can explore and process real possibilities to imagine and prepare for the future in a safe non-threatening situation (Thurow et al., 2023).

7.5 The Future Imagined

Of the existing body of research, rare examples stand out for different reasons. The first offers stark visions of the future. Linda Hassall's play *Dust* (2015) rearticulates the relationship between people and place by investigating the disappearance of the recognisable Australian landscape and raises themes of escalating temperatures, rising sea levels, disasters and a normalisation of these phenomena. Drawing on the Australian Gothic tradition, the play prophesises that humans will become dwarfed by non-human nature (Hassall, 2021). For her audiences, it is a wake-up call in the form of a contemporary performance investigation, as she introduces several hypothetical futures that provoke dialogue about what could happen if nothing is done about the climate emergency. Hassall attempts to force us to act and prepare for a different outcome.

Extraordinary work has also been done by Kris Verdonck's A Two Dogs Company, which presents large-scale scenarios rendering possible futures through a pared-back and poignant aesthetic (Thurow et al., 2023). One example is *PREY*, a music theatre piece initially inspired by the work of Australian ecofeminist Val Plumwood. Comprising "three solos by different generations of women, each has its own focus: text/language, song/music and dance/performance. With every solo, the tension between the human and landscape and/or performer and scenography, becomes more intense" (A Two Dogs Company, n.d.). After *PREY*'s frightening revelation that we are food and are belonging to a constitutive ecological cycle of life and death, it highlights how we nevertheless continue to destroy and undermine the foundations of life without displaying urgency towards the climate emergency. *PREY* makes us consider alienation from the environment and, in parallel, our alienation from our own bodies. Characteristic of Verdonck's oeuvre, the work becomes a nonlinear generation of dream-like fragments, each becoming more intimate and intense and giving us access to other perceptions and challenging our ideas on preparedness as it reflects "a world where humans are disappearing into the background" (A Two Dogs Company, n.d.) (Fig. 7.2).

Another important example is the series of experimental works by the Performing the Future team in the UK (2017–2018). With a focus on envisioning "positive futures in response to the anthropogenic climate and environmental change happening now" (Performing the Future, 2023), the group devised a series of workshops and digital performances to engage the public, combining scientific data, digital technology and audience participation. Of note are the paired works *The Prediction Machine* and *The Promises Machine* (2015), which were installed in locations across England during the project. *The Prediction Machine* tracked the weather at each specific location and predicted future climate change based on this and projected temperature increases between 2040 and 2050. It was an interactive digital work that combined weather data and video messages about climate change. Participants were asked to estimate weather variables by "embodied sensing" and to project changes into the future (Jacobs et al., 2019). The results were printed out as

Fig. 7.2 Photograph of *PREY*. (Image by Koen Broos)

predictions via "fortune telling machines". Participants were then invited to engage with *The Promises Machine* where they could "write their own promise for the future in response to the [prediction] they received" (ibid.). Participants filled in an online text box on an iPad. As project lead Rachel Jacobs points out: "The promises focus on opportunities for individuals to keep a commitment to something manageable within their own lives....They also explored opportunities to celebrate abundance and consider reassurance by reflecting on wishes for themselves and the world" (ibid.)

7.6 Preparing to Protect the Future

The most recognised artistic performance project embracing disaster preparedness in Australia has been *Refuge* (Arts House, 2016–2021), a world-first six-year programme based at the City of Melbourne's Arts House. Its goal was to address how art-based creativity might intervene directly and proactively in the climate crisis. In it, performance and creative artists were partnered "with experts from Australian Red Cross, State Emergency Services (SES), Emergency Management Victoria (EMV), University of Melbourne and local community organisations" (Wyatt et al., 2022). Together, they experimented with ideas around preparedness in the face of the growing climate crisis. In 2016, a local flood was imagined, and the North Melbourne Town Hall was turned into a relief centre for 24 hours, with actors and representatives of emergency and relief services collaborating to find solutions to problems. In 2017, five consecutive days over 40 °C were explored. In 2018, the theme was a pandemic event where the risk of contagion meant people might not be able to come together. Ironically, less than two years later, these imaginings were manifested in the COVID-19 setting. In 2019, displacement prompted by climate crisis was the focus, and in 2021, *Refuge* asked what would happen when these crises meet.

These performance-focused explorations considered practical ideas for implementation, striking a unique position, and the results outlined in the project's 2016 report highlight distinctive aspects of performance arts practices to support preparedness by creating spaces for enhanced and meaningful connection. They also generate aesthetic evocations of relevant emotions in safe ways. Additionally, the arts afford engagement of imagination and cultural practices, as exemplified in Indigenous practices, and we can explore how to combine with non-Indigenous approaches to create new kinds of resources and approaches. Arts organisations themselves also offer templates for relief and other kinds of centres, which bring community together (e.g. Yue et al., 2017).

Refuge has roots in "socially engaged art (Bourriaud, 2002), activist art (Sholette, 2017) and environmental art (Demos, 2016)" (Wyatt et al., 2022), but its performativity hovering between the representational and the real gives it a unique place. "The creative methods became the 'glue'...between diverse communities and practitioners to work together as a community" (ibid.). The approaches used strikingly contrasted to the "models and methods of conventional disaster management practice, offering an alternative to the 'template' approach" that was heavily critiqued by Steve Cameron from Emergency Management Victoria (ibid.).

In the realm of digital audio recording and sound art, Leah Barclay (2022) describes three case studies that draw on acoustic ecology and "eco-acoustics" to invite deep listening to changing ecosystems. Collaborating with local communities and traditional owners, the work centres Indigenous and place-based knowledges in exploring the impacts of climate change. One example, *River Listening*, engages audiences in soundwalks through digital app technology, virtual sound maps and geolocation, inviting them to listen deeply to compositions that have been created with local communities. Barclay describes how such approaches can provide "effective ways to engage communities in conservation and climate action" (2022).

Finally, in an exciting recent development, Jen Rae, an artist-researcher of Canadian Métis descent, and Claire Coleman, a Wirlomin-Noongar-Australian writer, have collaborated to create the Centre for Reworlding (C∞R; see Rae & Coleman, 2023): "A collective formed around our collaborative work intersecting arts, disaster risk reduction and resilience within the climate emergency context" (ibid.). This interdisciplinary art collective centres global Indigenous epistemologies to engage in experimental art practices of "reworlding". The project asks:

> What are the conversations that we aren't having now that might aid us, our loved ones, and our future ancestors?
>
> What are the skills and knowledges at the thresholds of being forever lost, overlooked, or undervalued that our future generations may need for survival?
>
> And what are we willing to give up and/or fight for in the greatest challenge facing humanity? Where do we begin? How will we reorganize?
>
> We begin by reworlding. (Rae & Coleman, 2023)

Such work represents an innovative speculative approach led by Indigenous artists and scholars in socially engaged performance arts. As the authors suggest, "Now more than ever, there is a role for arts and culture to lean into the tensions, to tell the unpalpable stories along with the rousing, and to ensure we have skin in the long game" (ibid.).

7.7 Conclusion

This chapter has demonstrated how performing arts projects have been used to deal with the aftermath of extreme weather events and highlighted their potential in developing preparedness for future ones. The examples, from a range of artistic and cultural traditions, show how different performing arts, for example, applied, experimental and digital, can reach diverse communities in presenting opportunities that not only help to re-enact crisis but also make sense of it in embodied and communal ways that build resilience and generate growth, supporting positive mental health and wellbeing benefits for those who have encountered climate disasters. The small body of existing work on disaster preparedness reveals how performance and allied embodied artistic practices can develop agency, empowerment, understanding and strategy for dynamic readiness in the face of disaster emergencies. Examples of performing arts works reveal the range of approaches including applied community-engaged work and large-scale imaginative pieces in which participants can become either immersed, alienated, connected/disconnected or provoked in ways that enable them to see future disaster scenarios in new ways.

Understanding the distinctiveness of the processes developed in *Refuge* in particular lays foundations for a new form of emergency preparedness that can enable relationships, ideas and practices to develop. Indeed, *Refuge* offers an excellent model, one that offers a series of bold steps in this increasingly important terrain, operating at the nexus of experimental art and disaster management to produce nuanced understandings and information and to foster preparedness. It was devised "as a knowledge broker to expand how [preparedness] might be understood and implemented" (Wyatt et al., 2022) and, implicitly, how resilience might be built. It has been recognised as having offered safe and open approaches, contained enough to guide, but never being didactic or formulaic, and embracing a range of people and scenarios. Rather than focusing on the outcomes such as the performance, it drew attention to "the dynamics of conversations, how ideas were transmitted and repeated, the circulation of stories and images, and the feelings that arose between and within people" (ibid.).

As we move forward, more work in the preparedness space is required to further understanding, knowledge and guiding practices. And now, in the 2020s, as technologies develop, the opportunities afforded by digital media through different forms of visualisation and sonic representation enable a greater range of performer and audience possibilities than ever previously experienced, as the former boundaries of geographical location or time zones disappear and more imaginative and expansive opportunities emerge.

References

A Two Dogs Company. (n.d.). PREY. *A Two Dogs Company*. https://t1p.de/lddr5. Accessed 6 Nov 2023.

American Psychological Association. (2023). *Resilience*. https://www.apa.org/topics/resilience. Accessed 27 Sept 2023.

Arts House. (2016–2021). *Refuge. Arts House*. https://t1p.de/tdzbj. Accessed 6 Nov 2023.

Australian Institute of Health and Welfare. (2020). *Australian bushfires 2019–2020: Exploring the short-term health impacts* (Cat. no. PHE 276). AIHW.

Barclay, L. (2022). Engaging communities in listening to ecosystems: Case studies from acoustic ecology research in Australia and Mexico. In S. Woodland & W. Vachon (Eds.), *Sonic engagement: The ethics and aesthetics of community engaged audio practice* (pp. 270–284). Routledge.

Bourriaud, N. (2002). *Relational aesthetics*. les presses du reel.

Cameron, L. (2020). Healthy Country, healthy people: Aboriginal embodied knowledge systems in human/nature interrelationships. *International Journal of Ecopsychology, 1*(1), 3.

Cecco, L. (2023). Wildfires turn Canada's vast forests from carbon sink into super-emitter. *The Guardian*. https://t1p.de/oh26e. Accessed 6 Nov 2023.

Collier, L. (2016). Growth after trauma. *Monitor on Psychology, 47*(10), 48.
Davidson, J. W. (2011). Musical participation: Expectations, experiences and outcomes. In J. W. Davidson & I. Deliège (Eds.), *Music and the mind* (pp. 65–87). OUP.
Davidson, J. W., & Garrido, S. (2014). *My life as a playlist*. UWA Publishing.
Davis, K. M. (2010). Music and the expressive arts with children experiencing trauma. *Journal of Creativity in Mental Health, 5*, 125–133.
Demos, T. J. (2016). *Decolonizing nature*. Sternberg.
Egan, C. J., & Quigley, M. C. (2015). Dancing earthquake science assists recovery from the Christchurch earthquakes. *Research in Dance Education, 16*(2), 161–183.
Fisher, N., & Talvé, A. (2011). *The role of arts in re-building community*. Arts Victoria.
Folke, C. (2016). Resilience (republished). *Ecology and Society, 21*(4), 44.
Folke, C., Carpenter, S., Elmqvist, T., Gunderson, L., Holling, C. S., & Walker, B. (2002). Resilience and sustainable development: Building adaptive capacity in a world of transformations. *Ambio, 31*, 437–440.
Freebody, K., & Finneran, M. (2021). Climate change—A battle for our hearts and minds. In K. Freebody & M. Finneran (Eds.), *Critical themes in drama: Social, cultural and political analysis* (pp. 153–166). Routledge.
Gallagher, K., Balt, C., Cardwell, N., & Valve, L. (2022). Arts-led, youth-driven methodology and social impact: 'Making what we need' in times of crisis. *International Journal of Qualitative Studies in Education*. https://doi.org/10.1080/09518398.2022.2098404
Garrido, S., Baker, F., Davidson, J. W., Moore, G., & Wasserman, S. (2015). Music and trauma: The relationship between music, personality, and coping style. *Frontiers in Psychology, 6*.
Gilmore, H. 2022. Anatomy of the Lismore disaster. *SMH*. https://t1p.de/bieuw. Accessed 6 Nov 2023.
Harms, L., Molyneaux, R. H., Nguyen, H., Pope, D., et al. (2023). Individual and community experiences of posttraumatic growth after disaster: 10 years after the Australian Black Saturday bushfires. *Psychological Trauma: Theory, Research, Practice & Policy*. Advance online publication.
Hassall, L. (2021). *Theatres of dust: Climate gothic analysis in contemporary Australian drama and performance landscapes*. Palgrave McMillian.
Hémond, Y., & Robert, B. (2012). Preparedness: The state of the art and future prospects. *Disaster Prevention & Management an International Journal, 21*(4), 404–417.
Heras, M., & Tàbara, J. D. (2014). Let's play transformations! Performative methods for sustainability. *Sustainability Science, 9*(3), 379–398.
Holling, C. S. (1973). Resilience and stability of ecological systems. *Annual Review of Ecology and Systematics, 4*(1), 1–23.
Horton, H. (2022). Summer wildfires increased fourfold in England in 2022. *The Guardian*. https://t1p.de/lagyj. Accessed 23 Sept 2023.
Igini, M. (2023). Summer 2023 was the hottest on record globally by a 'large margin', scientists say. *Earth.org*. https://t1p.de/vy4t3. Accessed 23 Sept 2023.
Jacobs, R., Abbott, F., Urquhart, L., & Price, D. (2019). Performing the future: An artist-led project engaging with risk, uncertainty and environmental change. *Journal of Risk Research, 22*(9), 1171–1185.
Kemp, L., Xu, C., Depledge, K., & Lenton, T. M. (2022). Climate endgame: Exploring catastrophic climate change scenarios. *Earth, Atmospheric & Planetary Sciences, 119*(34), e2108146119.
Kendra, J. M., & Wachtendorf, T. (2003). Elements of resilience after the World Trade Center disaster: Reconstituting NYC's emergency operations centre. *Disasters, 21*, 37–53.
Kurth, C., & Pinkala, P. (2022). Eco-anxiety: What is it? *Frontiers and Psychology, 13*.
Lazo, J., Bostrom, A., Morss, R., Demuth, J., & Lazrus, H. (2015). Factors affecting hurricane evacuation intentions. *Risk Analysis, 35*(10), 1837–1857.
López-Marrero, T., & Tschakert, P. (2011). From theory to practice: Building more resilient communities in flood-prone areas. *Environment and Urbanization, 23*(1), 229–249.

Madrigano, J., Chandra, A., Costigan, T., & Acosta, J. D. (2017). Beyond disaster preparedness: Building a resilience-oriented workforce for the future. *International Journal of Environmental Research & Public Health, 14*(12), 1563.
McFerran, K., & Teggelove, K. (2011). Music therapy with young people in schools: After the Black Saturday fires. *Voices, 11*.
Mullins, G. (2021). *Firestorm. Battling super-charged natural disasters*. Penguin.
O'Connor, P. (2013). TIE: The pedagogic as the aesthetic in a crumbling world. In A. Jackson & C. Vine (Eds.), *Learning through theatre: The changing face of theatre in education 3* (pp. 305–320). Routledge.
Peek, L., Tobin-Gurley, J., Cox, R., Scannell, L., Fletcher, S., & Heykoop, C. (2016). Engaging youth in post-disaster research: Lessons learned from a creative methods approach. *Gateways: International Journal of Community Research and Engagement, 9*(1), 89–112.
Performing the Future. (2023). *About*. https://performingthefuture.net/about/. Accessed 7 Dec 2023.
Rae, J., & Coleman, C. G. (2023). Reworlding: Speculative futuring in the endtimes, in the everywhen. In S. Lammes, K. Jungnickel, L. Hjorth, & J. Rae (Eds.), *Failurists: When things go awry* (pp. 69–78). Institute of Network Cultures.
Sholette, G. (2017). *Delirium and resistance: Activist art and the crisis of capitalism*. Pluto.
Stephens, A. (2022). Loss and hope: How Lismore's arts community is recovering from the floods. *Art Guide Australia*. https://t1p.de/q2hys. Accessed 1 Nov 2023.
Tedeschi, R. G., & Moore, B. A. (2021). Posttraumatic growth as an integrative therapeutic philosophy. *Journal of Psychotherapy Integration, 31*(2), 180–194.
Thurow, S., Grehan, H., & Davidson, J. W. (Oct 2023). Performing Arts and the climate emergency. Horizon-scanning the futures of practice and scholarship. *Australasian Drama Studies, 83*, pp. 12-37.
United Nations Office for Disaster Risk Reduction. (2022). *Global assessment report on disaster risk reduction*. UNDRR.
Williams, H. (2010). How to make a drama out of a disaster. *The Independent*. https://t1p.de/66u5i. Accessed 1 Nov 2023.
Woodland, S., Hassall, L., & Kennedy-Borissow, A. (2023). *Embodying resistance and survival: How youth theatre and drama can address eco-anxiety and promote recovery in disaster affected communities*. University of Melbourne.
Wyatt, D., Pfefferkorn, J., & Papastergiadis, N. (2022). *Finding refuge: Art, emergency and "imagining the unimaginable" in the resilient city*. University of Melbourne.
Yue, A., Trimboli, D., & Di Biase, T. (2017). *Refuge 2016 evaluation report*. University of Melbourne & Arts House.

Chapter 8
Iconographies of Climate Catastrophe: The Representation of Climate Change in Art and Film

Charles Green, Belinda Smaill, and Seán Cubitt

Abstract This chapter reviews visual representations of climate in art and film across the last few decades, exploring shifting artistic, cinematic, televisual and narrative practices that have more recently shaped the communication of the climate emergency. It explores the iconographies that artists and filmmakers have used in the shift from representing contemporaneous and local environmental challenges to depicting the future consequences of climate warming, including widespread biospheric change. The authors sketch the shifting screen media formulations that imagine climate catastrophes, observing how both documentary film, television and contemporary art draw on popular and professional media practices, including news formats and visual effects.

Keywords Cinematic · Climate change · Climate warming · Contemporary art · Data visualisation · Documentary · Film · Imaging · Screen media · Video installation

8.1 Introduction

The subject of this chapter is the visual representation in art and film of climates and the biosphere across the last six decades. The chapter locates the iconographies that artists and filmmakers used in the shift from representing environmental threats to

C. Green (✉) · S. Cubitt
School of Culture & Communication, The University of Melbourne (UM), Parkville, VIC, Australia
e-mail: cdgreen@unimelb.edu.au; sean.cubitt@unimelb.edu.au

B. Smaill
School of Media, Film & Journalism, Monash University, Caulfield East, VIC, Australia
e-mail: belinda.smaill@monash.edu

D. Del Favero et al. (eds.), *Climate Disaster Preparedness*, Arts, Research, Innovation and Society, https://doi.org/10.1007/978-3-031-56114-6_8

depicting the future consequences of climate warming and, specifically, of widespread biospheric change. Our proposition is that there has been a distinct change in the representation of climates and the natural environment shaped by artists' intentions to depict the climate emergency.

Visual art produces new models of the world with the aim of gaining public currency. An emerging literature on art and climate warming (e.g. Fowkes & Fowkes, 2022) has already focused on establishing a typology of iconographies of climate warming. We sidestep definitions and think historically, focusing on a succession of visual artists and film practitioners, identifying their ideas about those issues rather than imagining that they have been invariant or, even more reductively, imagining that eco-critique would only appear in climate-themed works. For instead, we find the climate crisis coded across a kaleidoscope of art and film, including in works within which no actual representation of climate warming is foregrounded. But the climate emergency is still there. In this respect, we chart the history and immediate pre-history of climate awareness in art, film and television.

In other words, we will suggest that the iconography of environmentalism was substantially superseded by the iconography of climate warming. The duration of this shift coincides with the period of contemporaneity (i.e. since the 1960s), which Terry Smith (2009) comprehensively defined in art. The period has now been long enough to periodise representations and iconologies of climate.

So, we sketch here a history of climate change art in the period of contemporaneity, a period of transition from imaging environmental loss to imaging climate warming when artists and filmmakers approached climate and environment with similarly changing forms and visualisations—similar reimagining—on their minds, if not in their intentions.

As a result, a newer iconology of climate warming was imagined between the 1960s and the 2020s that often plundered and retrofitted the technologies and forms that the directly preceding generation had taken up on account of the affective and technological immediacy of new media and film. Early in this period, we see the hegemony of colour photography and video (and, off to the side, niche innovations such as Cinerama, the 1960s precursor to IMAX cinemas) and the efflorescence of then new genres such as broadcast television's environmentalist documentaries. Both faded and at the same time were transformed into signifiers of the now-lost past. To track these changes across three sub-periods, we have borrowed descriptions of the stages of grieving to mark 1960–1980 as a period of denial and bargaining, 1980–2000 as grief and anger and 2000 to now as depression and—perhaps optimistically—understanding (Kübler-Ross, 1973).

Our account draws on prior insights: art theorist John Berger (1972) read the whole political economy of enclosure and looming industrialisation in the eighteenth-century England into Gainsborough's *Portrait of Mr and Mrs Andrews* (c.1750). Nicholas Mirzoeff's 2016 essay on the Anthropocene sublime in Turner and Monet's London paintings was in turn preceded by Michel Serres's famous early 1990s lectures (1997) on volcano-tinted sunsets in the same nineteenth-century paintings. The long history of climate change, artistic responses and critical acknowledgement is implicit in what follows.

8.2 Period 1 (1960–1980): Science Aesthetics—Denial and Bargaining

A new aesthetics of science, environment and landscape emerges in the 1960s, helped by illustrated books like Gyorgy Kepes's *The New Landscape in Art and Science*. Like Kepes, Charles and Ray Eames's film *Powers of Ten* (1968–1977), commissioned by IBM, was a fruit of a wider practice they referred to as "a natural overlap with several governmental agencies" (Lipstadt, 2005) and drew on postwar cybernetics and then-emerging Earth systems science. The peak of systems theory coincided with global transitions from modernism to contemporaneity centred around the year 1962, which also saw the publication of US environmentalist Rachel Carson's groundbreaking book, *Silent Spring*, and Australian environmentalist novelist Nancy Cato's visionary *But Still the Stream: A Novel of the Murray River*. The positivist, glossy, blue-chip amplitude of American nature documentaries—the "*National Geographic* aesthetic"—was taken up as a recognisable set of tropes in Hollywood genre films such as *How the West Was Won* (1962), produced as feature-length spectaculars in Cinerama, a short-lived immersive format requiring its own cinemas (in Melbourne, the remodelled Regent Theatre) kitted out with multiple projectors. A predecessor of IMAX, Cinerama encouraged elegiac but positive environmentalism: the poetic paradigm of pristine nature even when endangered by rapacious capitalism. The same *National Geographic* environmentalism was embodied in Australia in Olegas Truchanas's 1971 colour photographs of the about-to-be-dammed Lake Pedder in Tasmania, all-white sands and crystal clear waters. Many countercultural versions of this Walden-like, spiritually elevating vision of nature appeared at the time, including Stan Brakhage's well-known alternative films such as his psychedelic *Dog Star Man* (1961–64).

The 1970s brought an intensification in environmental activism within art. While there were differences in emphasis across international locations, the 1960s environmentalism emphasised wilderness preservation. Now, there was a growing fight against air and water pollution and pesticides. The 1970s artists linked environmentalism to the end of the modernism that had dominated art since the 1860s. This was the period of minimalist sculptor Carl André's *Scatter Piece* (1967), one of the many exhibitions of waste and detritus thrown across art gallery floors as conceptualist art, and of Robert Smithson's *Spiral Jetty* (1970), land art of rock and earth bulldozed out from Rozel Point in the north arm of the Great Salt Lake in Utah. Artists saw environmental decay as an artistic trope as well as a cause. Robert Smithson (1968) equated environmental decay with the undermining of representation and felt sheer excitement at seeing entropy and environmental decay in action. He was not necessarily the friend of environmental campaigners, and they reviled him for *Island of Broken Glass*, his never-realised 1970 attempt to cover an island off the coast of Vancouver in 100 tons of glass shards. He disfigured photographs of wilderness and made arrangements in art galleries of shattered glass, rock and sand, opening his works (and their viewers) to imagining geological force. Rather than depicting environmental damage, he saw local landscapes in global contexts and in

Fig. 8.1 Robert Smithson's *Spiral Jetty*. (Image by Netherzone)

geological durations. Other 1970s art collaged together images in which environmental loss was one part of an unstable mix of subjects, experiences and forms: wild tourism, science fiction, alternative film, the counterculture, ecological theory, different spiritualities and cosmologies. Early computing and jet flight organised artists' experiences of all these (Fig. 8.1).

By 1970, a collective environmental politics induced a shift from an idea of environment as the local effect of elemental forces to the realisation of a planetary and future-oriented problem whose scale and urgency demanded a new episteme of representation. Environmentalist sense of global connection saw the establishment of Earth Day in the United States and was visualised in the famous "Blue Marble" photograph taken by the Apollo 17 mission in 1972. A related illustration featured on the cover of Stewart Brand's *Whole Earth Catalog* (1968–1972), another systems-inspired publication, as important for its era as Kepes's book two decades earlier, and an important starting point for the emerging countercultural "green" consumerism. Ursula Heise (2008) notes that this photograph contributed to the promotion of a "holistic understanding of ecological connectedness", which built not only on new thinking about the environment and earth systems but also suggested "a unified and balanced world" in the face of the divisions of Cold War politics and the counterculture's alternative institution-building of the 1960s. It took another decade, however, to connect planetary consciousness to the problem of climate change. While the issue of global warming was included in the UN environmental framework by the early 1970s (Warde et al., 2021), the same period as the formation of the US Environmental Protection Agency, scientific consensus was only just forming, and public opinion was slower to mobilise than either artists or activists.

The connection between global and local established in the 1970s, not least due to activism against nuclear power and testing, was taken up by artists as a problem of communication that cinematic aesthetics could solve. Paul Ehrlich's influential book *The Population Bomb* (1968) and its Malthusian prediction marked art house films such as Michelangelo Antonioni's *Zabriskie Point* (1970), with its naked hippies wandering in unexplained post-catastrophic desert landscapes, while a sleek, Buckminster Fuller–like residence explodes in slow motion to the white-noise crescendo of Pink Floyd. Its tremors spread to popular films such as *Soylent Green* (1973; see Murray & Heumann, 2009), which ends with the film's hero in tears at the spectacle of green forests and wild spaces, screened to soothe his euthanised mentor in a scene whose pathos arises from both the loss of nature and the inadequacy of film to capture it. The theme of media's inadequate affect would reappear in postmodern artist Richard Prince's series *Untitled Cowboys* (1980–1992), pastiches of rural machismo and wild horses appropriated from US advertising. Like their sources, Prince's mural-sized photographs reduced landscape to the backdrops of an elegiac dream. Population anxiety and anti-nuclear activism were leading image makers to nihilism and the limits of visualisation. Even greater problems of invisibility plagued attempts to move climate change beyond the realm of science communication. Unlike weather, which everyone experiences every day, climate change is an abstraction from the collection and comparison of data. In a symptomatic statement, land artist Walter de Maria (1980) wrote of his 1977 high-desert installation, *Lightning Field*, that "The invisible is real".

At the same time, non-narrative documentaries, including feature-length films, appeared, which aestheticised and narrated but also doubted and destabilised science, for instance, Bay Area artist Bruce Connor's hallucinatory repurposing and re-editing of found US Government nuclear test documentation *Crossroads* (1976) and Michael Glasheen's avant-garde video *Uluru: Mythology of the Dreamtime* (1978). Another work influenced by US architectural theorist Buckminster Fuller, *Uluru*, used "time-lapse photography, superimpositions, [early-generation video editing] and rapid montage" (Barker & Green, 2010) to portray the central Australian rock monolith. As a marker of the currency of such depictions, *Uluru* was screened to sizeable audiences in Melbourne and Sydney.

By the late 1970s, science reporting was bringing climate change to mainstream audiences. But the causes and effects of climate change remained in the realm of scientific research and communication. Science television was an early site for grappling with the challenge of this new episteme. A rare example was the television documentary *A Change of Climate* (1977), a collaboration between a national public broadcaster, the Australian Broadcasting Corporation (ABC), and Australia's peak scientific research organisation, the Commonwealth Scientific and Industrial Research Organisation (CSIRO). This 46-minute film, based on the 1977 Australasian Climatology Conference, was a surprisingly comprehensive snapshot of climate science at that moment in time. It brought the established lexicon of science films to the new problem of climate change: improvised, opportunistic interviews and scientists in dialogue with one another, clips of scientists in the field collecting earth and ice cores, launching weather balloons and working on space stations. Most significantly for our discussion, the presentation of data was crucial.

In 1977, scientists were debating the question of whether the planet was on track for another ice age or for a warming climate. *A Change of Climate* offered information, combining authoritative expert voices with graphic visual representations, both lined up into a narrative. This is not as obvious as it looks: for instance, the data itself remained as indecipherable as the technology, both commonly shared signifiers of benevolent expertise. The objects of technological progress loaned *A Change of Climate*'s narrative its legitimacy rather than specific meaning. Scientists pointed to maps and satellite photographs. Computer screens flickered, and large, cumbersome printers churned out reams of paper. With a focus on Australian research projects, the scientists at the conference offered the veneer of internationalism even though a planetary perspective was still minimised. The film oscillated between a possible ice age thousands of years hence and predicting climate warming, worrying whether it should be communicated given the current evidence and, if so, how. One prescient meteorologist predicted that the climate would warm by half a degree centigrade by the end of the century, noting that the ramifications of this would be immense. In sum, *A Change of Climate* shows us a wide range of representational devices from this first period of representing climate change, many of which, such as the tropes of ice cores and polar regions, can still be seen in contemporary climate documentaries or art.

8.3 Period 2 (1980–2000): Environmental Ruin and Nuclear Decay—Anger and Grief

Soviet film director Andrei Tarkovsky's feature *Stalker* (1979) attacked settled assumptions about scientific positivism and a global world order that had come before, alienating Soviet-era cultural bureaucrats and depicting environmental and spiritual ruin. By contrast, the uncannily similar plot in Steven Spielberg's *Close Encounters of the Third Kind* (1977) initially portrayed the kaleidoscopic global interconnection of vast new mysterious phenomena but quickly retreated into scientism, the respectful portrayal of dedicated scientists and their well-intentioned but over-secretive government and military. *Stalker* presaged a shift into the second phase of imagining climate change, a phase in which the tropes of nuclear annihilation and ruin were imported into environmentalist art and film. An agonisingly slow, science fiction film shot in subtle, shifting monochrome tones, *Stalker* documents the painful, illegal trek of a group of amateur explorers, including The Scientist and The Writer, led by a young boy, The Stalker, across a mysterious and dangerous ruined post-industrial Zone. Its images of irredeemable ecological decay indelibly changed many environmentalist artists' sensibilities. Composed of telling pauses where the camera focuses on the actors' tormented faces, nested within long sequences raking slowly across decaying underwater collages, *Stalker* offered a panoramic view of environmental decay akin to Minimalism in its strict refusal of narrative exposition. At the same time, it proliferated suggestive environmental allegories, akin to the Land Art we outlined earlier. It was not alone: in the United States, Godfrey Reggio's feature-length experimental film, *Koyaanisqatsi: Life Out of Balance* (1982, scored by minimalist composer Philip Glass), was wildly successful, attracting large art house cinema crowds to gape at its spectacular, slow tracking shots and time-lapse photography of vast landscapes alternating with accelerated static camera sequences of crowds swarming across cities. Reggio repeatedly punctuated these sequences with an enigmatic, slow-motion collage of explosions gathered, like Connor's *Crossroads* a few years before, from found footage, manipulated and slowed down on the editing desk: the superimposed shapes of falling buildings, a failed rocket launch, nuclear tests.

Neither *Koyaanisqatsi* nor *Stalker* ended with an environmentalist punchline nor with a discernible anti-nuclear message. *Stalker's* little band of illicit explorers repeatedly hint that the Zone appeared just after a mysterious visit from outer space. Tarkovsky's tortured wanderers did not arrive at redemption though they did arrive at their destination: an innermost room, the ruined repository of immanent truth. *Stalker* did not move beyond this simple plot nor offer any character development, focused instead on intricate details of terrible ruin. Tarkovsky's profoundly anachronistic dystopic film remains an underground cult among contemporary artists, predicting the aesthetic of the third wave of climate change art.

Nuclear proliferation and the threat of nuclear war dominated the imagery of climate issues, shaping the 1980s science communication but also avant-garde experimentation. In Australia, artist Bonita Ely presented *Jabiluka* UO_2 (1979), a ritualistic anti-uranium performance inspired by campaigns for Indigenous land rights and by her visit to the edge of the Jabiluka uranium mine, which had been excised from the Kakadu National Park against the wishes of its traditional Aboriginal stewards. *Jabiluka* UO_2 was performed in Melbourne in 1979 and then at the halcyon 1980 performance festival, ACT 2, in Canberra. Ely, dressed in a white boilersuit, erected an elaborate sandcastle from coloured earths and then arranged a spiral of flammable straw across the adjacent lawn. Her male assistant traced a rigid, straight path in white lime across the field towards her, eventually scattering the sand construction. Simultaneously, Ely set fire to the straw, etching a primal spiral design onto the grass field.

By the early 1980s, the CSIRO's Film Unit had produced more educational science films either mentioning or fully highlighting the threat of climate change. For example, *The Coal Question* (1982), directed by Russell Porter, looked directly at coal processing and social responsibility, paying brief attention to explaining the problem of the "greenhouse effect". Then, in 1984 for the same unit, Porter directed *What to do about* CO_2*?* It was not until this period that the greenhouse effect became part of the popular cultural imagination. The broadcast lexicon of science and technology as signifiers of authority and expertise persisted. However, they were surpassed in the early 2000s by data visualisation, which gave the impression of direct presentation of meanings, as opposed to appearances.

By the mid-1980s, as awareness of climate science was growing in Australia, nuclear power, testing and uranium mining were well established as primary concerns in the environmental movement and preoccupied the attention of image makers and artists to an even greater extent. This must be remembered since it is at odds with artists' climate priorities today. The feature-length documentary film *Half Life: A Parable for the Nuclear Age* (1985) was a landmark exploration of the impact of nuclear testing in the Pacific. At the same time, the Indigenous communities of the areas surrounding the testing sites, in the APY homelands, began their ascent to art stardom, often incorporating into paintings and sculptures their experiences of dispossession through nuclear tests and sickness from radiation.

In the late 1980s, the problem of the greenhouse effect and the depletion of the Earth's ozone layer (a problem first discovered in 1983) came to the forefront of public debate and media attention as combined threats to the planetary ecosystem. A headline in *Time Magazine* in October 1987 read: "The Heat is on: Chemical Wastes Spewed into the Air Threaten the Earth's Climate". In 1988, the Intergovernmental Panel on Climate Change (IPCC) was established. In 1990, Earth Day became an international event, and the Earth Summit (also known as the UN Conference on Environment and Development) took place in 1992. The 1990s, which also witnessed the emergence of the Internet, was an optimistic decade for environmental action. The new policy infrastructure recognised the full range of environmental issues and the urgent need for solutions. Documentary filmmakers produced films on environmental issues, and environmental film festivals germinated around the world. The first of these was the Tokyo Global Environmental Film Festival, which began in 1992. As Charles Musser writes, these festivals "played a crucial role in the construction of the genre's identity" (2014) bringing a new, high public profile environmental role for documentary film. In a pivotal combination of environmental governance and art, UNESCO commissioned Yann Arthus-Bertrand in 1994 to create an unprecedented work of documentary photography from the air. From a journey across 5 continents and 60 nations, Arthus-Bertrand produced an inventory of landscape images that elicited grief and a sense of what was being lost in the face of environmental degradation, including climate warming. By the early 2000s, the exhibition, book and film that resulted from Arthus-Bertrand's project had reached an audience of millions. On the heels of the 1990s reinvigoration of the documentary form abetted by new light portable digital cameras and high-quality digital editing software, there was widespread public acclaim for films like Agnes Varda's *The Gleaners and I* (2000) documenting gleaners (i.e. the collectors of remainders from harvests and refuse collections) and *Darwin's Nightmare* (2004, dir. Hubert Sauper) on the destruction caused by the Nile perch, introduced to Lake Victoria by humans.

8.4 Period 3 (2000–2023): Climate Warming—Depression and Understanding

One of the most notable shifts in the early 2000s was the inclusion of environmental messaging in the media's most visible television presentations about the natural world. The BBC's Natural History Unit's distinctive, high-budget, landmark style was first developed by Sir David Attenborough in the 1970s with *Life on Earth* (1979). His commercial model for presenting natural history on television, even publicly funded television, evolved more slowly and more cautiously than we might remember. Both high-cost and highly successful in gaining audience reach, Attenborough's documentaries took a critical role in offering an international visual imaginary for nature, but BBC investment in "prestige productions" came with guidance against political controversy, and his numerous series avoided discussion of anthropogenic environmental impacts until well into the 2000s. Morgan Richards's assessment of the "greening" of natural history documentary placed Attenborough's first unequivocal statement about the veracity of climate warming in 2006 (2013). In 2000, the BBC Natural History Unit allowed Attenborough to present *State of the Planet* (2000), a three-part series that described a series of threats to habitat and ecosystems. The series had less investment and reached smaller audiences than Attenborough's blockbuster prestige series. Nor did it foreground climate change. The final episode ends with a call to action but eschewed directly naming the elephant in the room, catastrophic climate warming:

> The future of life on earth depends on our ability to take action. Many individuals are doing what they can, but real success can only come if there's a change in our societies and our economics and in our politics. I've been lucky in my lifetime to see some of the greatest spectacles that the natural world has to offer. Surely, we have a responsibility to leave for future generations a planet that is healthy, inhabitable by all species. (Attenborough, 2000)

The series failed, like almost all mainstream film, to link climate warming to the global diaspora of refugees and displaced peoples or to wars and conflicts around the planet.

By the time *Frozen Planet* was released in 2011, a profound shift had occurred to the degree that the final episode in the series was now directly devoted to the problem of climate change and its impact on sea ice, glaciers, ice shelves and polar habitats. Though broadcast in the United Kingdom, controversy erupted when that last episode was made optional for syndication for the United States. US Discovery Channel eventually added the seventh episode to their broadcast schedule.

At the same moment, contemporary artists were reintroducing the nature documentary into the art world but were quoting, exploding and shattering that genre's images and its self-conscious narrative continuity. Prominent LA artist Doug Aitken's multi-channel immersive video installations had been influenced by earlier periods of environmentalist cinema, including multi-screen world exposition technology, not least Cinerama—both in terms of technology and hybridity of genre and narrative. Aitken's work implies that he was intensely aware of the successive phases of those early documentaries about both environmental majesty and environmental decay. His video, *New Ocean* (2001), consisted of joined images, right-way-up and upside-down, in kaleidoscopic reconstructions of watery landscapes from Brazil, Alaska and California. Across an array of scrim-covered projection screens, water flowed upwards, and cliffs were split and recombined like Rorschach tests in an IMAX delirium. "Aitken himself was so concerned to emphasise precisely this hyperactive non-linearity that he interviewed other artists and film directors on the subject, publishing the resulting collage of voices as…a book, *Broken Screen*" in 2006 (Green, 2008). His works resembled unfolding explosions and their aftermaths, in many ways building on images evoked in the 1970s cinema with films such as *Zabriskie Point* and *Soylent Green.* Aitken, like other high-profile artists participating in biennials, triennials and large survey exhibitions around the globe that had come to define the vast industry of contemporary art, excluded data visualisation and didactic exposition from his art. Yet, the next generation of artists would explore its aesthetic possibilities.

The peak moment for data visualisation in climate change films, emerging from the 1970s and 1980s in documentaries such as *A Change of Climate*, was Al Gore's celebrated and controversial feature, *An Inconvenient Truth* (2006), with its climactic presentation of the famous "hockey stick" graph. *An Inconvenient Truth* was exploring a new aesthetics in which compelling, dramatic images, didactic exposition and Eisensteinian montage were all harnessed together to amplify the importance of global climate change. This issue now superseded all other ecological agendas and stood alone as the foremost problem for the Earth Sciences. The film's willingness to delve into data was about to be taken up by visual artists.

An Inconvenient Truth premiered at the Sundance Film Festival alongside another important documentary: Chris Paine's *Who Killed the Electric Car* (2006). Both films were part of a wider movement documenting the role of corporations and governments in perpetuating powerful, fossil-fuelled capitalism in the face of the climate crisis. Yet more important environmentalist films were released in 2006, Jennifer Baichwal's *Manufactured Landscapes* and HBO's *Too Hot Not to Handle*, which documented the effects of global warming. Filmmaking on environmental issues intensified over the next few years. In 2009, the Sundance Film Festival premiered six environmentalist documentaries that included *The Cove*, *Crude*, *Earth Days* and *The End of the Line*. As Musser (2014) observes, Sundance 2009 was referred to as the "Green Festival", the crest of the wave of environmental filmmaking produced during the last year of the George W. Bush administration. In other nations, including the United Kingdom, where public broadcasting and the commissioning of independent documentary are more tightly linked, philanthropic funding was on the rise. This led to greater investment in social issues and, in turn, in funding for environmentalist documentaries. The "Good Pitch" initiative was started in 2008, piloting a funding infrastructure for private investment. "Good Pitch Australia" followed in 2015 and led to a minor but marked national rise in documentary filmmaking committed to environmental issues. The highest-profile film to emerge from Good Pitch Australia was *2040* (2019).

By the 2010s, neither the news media nor artists could take the weather or nature for granted. At the heart of the coal-themed European biennial of contemporary visual art, Manifesta 9, held for its 2012 edition in the semi-derelict Belgian lignite-mining town of Genk, was a succession of darkened rooms lined with drawings, film projections and prints curated by a prominent art historian of modernism, Dawn Ades. It traced the entropic history of coal mining in art from the start of the Industrial Revolution through its collapse during the 1980s European deindustrialisation to the emergence of climate warming as a social issue. In film as in art, climate warming's iconography rethought and assimilated older, competing apocalyptic fears (pollution, nuclear holocaust, pandemic) into the fear and dread of climate catastrophe. Ian Cheng's post-apocalyptic *Emissary in the Squat of Gods* (2015) ran in real time on a games engine to portray a world of aimless, broken survivors in a devastated world, while Tabita Rezaire's *Deep Down Tidal* (2017) tracked the traces of lost Black wisdom and history in the deep oceans. Both works emphasised environments as witnesses, guardians and active agents in the making of experience, echoing Walter de Maria's note that "The land is not the setting for the work but a part of the work" (1980).

The same themes reached quite separate culminations and an apotheosis in the immersive video installations of Black British filmmaker John Akomfrah and of Irish photojournalist-turned-filmmaker Richard Mosse. Both have become key figures in international contemporary art. Both emerged from professional documentary backgrounds. Akomfrah was a founding member of British Black Audio Film Collective, launched in 1982, which made socially activist documentary films from an explicitly Black perspective. From the 2010s on, Akomfrah's immersive video installations focused on climate warming. *Vertigo Sea* (2015) reimagined the iconography of the natural world through the lens of memorialisation and grief. In *Vertigo Sea*, the broken narratives are drawn from archival footage, including bloody, gruesome whaling; long slowly drifting aerial sequences of wilderness, wild animals and migrating birds; and performances by actors in period costumes "who gaze silently at landscapes threatened or already blighted by human progress" (O'Hagan, 2017). His next work, *Purple* (2017), was, in his words, "a response to [the] Anthropocene" (qtd. in O'Hagan, 2017). Akomfrah conveys a conception of a bleak colonial past haunting the present by the intertwined history of imperial expansion, slavery, colonisation, war and conflict, all now linked to climate warming. Environmental destruction and historical displacement—the mass mortality of native Americans and the industrialisation of African slavery after Columbus—kickstarted capital and bonded it permanently to climate warming and migration (Baucom, 2020; Mignolo, 2009). The global flow of the decolonised, of refugees and of displaced asylum seekers, driven by climate-related famines, water shortages and wars, only reverse European expansion since the fifteenth century.

Two questions have haunted this chapter. What impact did this art and film practice have? What use were they in communicating climate warming's impact? Initially imagined from scientific and emotional heritage of environmentalism, climate warming was reimagined in relation to human and non-human species' displacement. Data was first visualised as an aesthetic effect but later patiently presented and unpacked to inform and educate. Widescreen cinema first celebrated winning the race to colonise. And then its successor, immersive video, mourned colonisation's dark articulation with environmental destruction in a grief-driven displacement aesthetics. First Nations peoples and refugees were questioning both the ethics and the efficacy of this reimagining and of how displacement was depicted, no matter what the immersive affect or artistic quality of environmentalism and the emergent displacement aesthetics that linked the loss of landscape to human flows across borders and to climate warming. They were sceptical about environmentalism and about the landscape of contemporary art that largely excluded them. They were sceptical about reimagining the displacement of humanity and non-human species through sweeping panoramic abstractions. That scepticism is encapsulated in the demand, Nothing About Us Without Us.

More collaborative practices therefore became increasingly important. In the 1970s, public artists Helen Mayer Harrison and Newton Harrison employed "fieldwork techniques from sociology, such as interviews and community [consultations to gather] vast amounts of information about environmentally threatened sites. From this [information], they made series of collages, photomurals and books combining enlargements of aerial photographs, maps and illustrative drawings with evocative texts, captions, and detailed descriptions of ecological strategies were designed to heal degraded or damaged regions" (Green, 2001).

Forensic Architecture, an artistic method developed by the collective of the same name based at Goldsmiths College in London, continue this tradition, analysing and presenting forensic evidence for legal and political processes, re-presented as art. Forensic Architecture investigate human rights and environmental violations in partnership with grassroots activists, legal teams, international NGOs and media organisations. Their "art" "involves open-source investigations, the construction of digital and physical models, 3D animations, Virtual Reality environments and cartographic platforms" (Lorca Macchiavelli, 2019). Their projects have involved state-of-the-art data visualisations of information gleaned from "photographs, videos, audio files and [verbal] testimonies [to] reconstruct and analyse violent events" (ibid.). In *Ecocide in Indonesia* (2017), Forensic Architecture collaborated with local activists and scientific observers, and they pinpointed responsibility for environmental arson in Kalimantan and Sumatra, with a visual effect poised between horror and cold legal work. They began to demonstrate the potential of reimagining environments through network technologies, in effect criticising the politics of previous climate data visualisations and artistic representations since *An Inconvenient Truth* for avoiding blaming specific economic or political agents and instead portraying the Earth as victim. *Ecocide in Indonesia* may have looked poetic but was in fact unambiguously practical, suggesting concrete, socially viable environmental strategies. Forensic Architecture demonstrated a form of useful art, produced in collaboration with local and Indigenous peoples, freeing systems theory of its subordination to control in the 1960s cybernetics and its legacy and articulating it with ordinary people's pictures, sounds and words. Transforming "found" media also transformed the ordinariness of witnesses. It transformed art's normative and apolitical assumptions into action plans and undercut the inherited authority of expertise (Fig. 8.2).

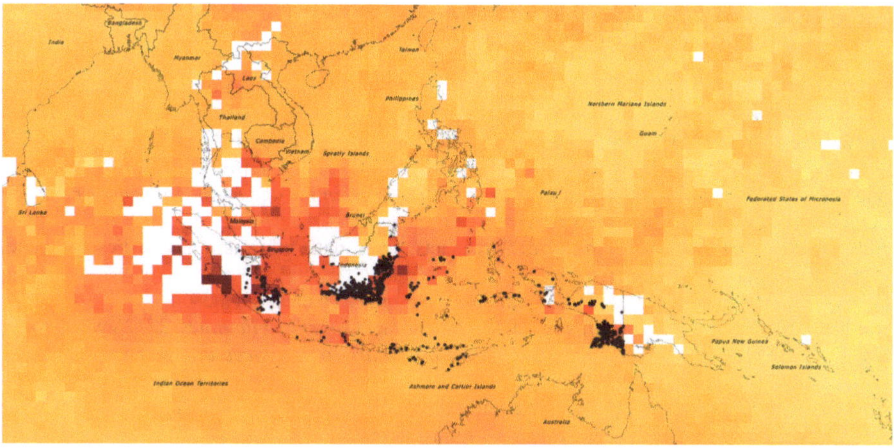

Fig. 8.2 "Carbon Cloud and the sources of fire, 2015." *From Ecocide in Indonesia*, 2017 © Forensic Architecture

8.5 Conclusion

Climate emergency and climate warming are synthetic and cultural concepts. They have an art and film history coincident with the period of globalised contemporaneity. A genealogy of forms through which artists reimagined the climate emergency shows how they dramatised broken narratives and excavated tropes ranging back to Romanticism, including the pathos of victimisation, poetic and elegiac visions of nature and images of dystopic annihilation. Our period was formed by struggles between evocation and communication; dawning realisation of the continuity of colonialism, capitalism, climate and human displacement; and a struggle between dread of unforeseen catastrophes and fear of technocratic attempts to control them. Both art and media historians have a tendency to periodise: so now, in the third decade of the twenty-first century, we should periodise environmentalism.

We have outlined three periods wherein the iconography of nature remained similar, but its denotations shifted profoundly, as did artists' technologies. Once upon a time, the threat of nuclear war and the end of the world might be denoted by an audio recording of Geiger counters. In the 2020s, the same clicking conjures up Fukushima, whose catastrophic flooding and subsequent release of contaminated water evokes not just geological or nuclear crises but also reminded the world of climate warming's coming sea rises. Our chapter showed how climate iconography reimagined and assimilated older, competing fears of apocalypse: of nuclear holocaust, of industrial pollution, of Indigenous dispossession. The connections that artists and filmmakers made to older art and earlier matter-of-fact documentary ways of seeing led inexorably to their engagement with the techniques, ethics and politics of data management and visualisation. The contemporary awareness of ecological impact has become, above all, a very political awareness, largely due to the artists and filmmakers of this period.

References

Aitken, D. (2006). *Broken screen: 26 conversations with Doug Aitken expanding the image, breaking the narrative*. Turtleback.
Attenborough, D. (2000). *State of the planet*. BBC Natural History Unit.
Barker, H., & Green, C. (2010). No more provincialism: Art & text. *EMAJ, 1*(5), 1–30.
Baucom, I. (2020). *History 4° Celsius: Search for a method in the age of the Anthropocene*. DUP.
Berger, J. (1972). *Ways of seeing*. Penguin.
de Maria, W. (1980). The lightning field: Some facts, notes, data, information, statistics, and statements. *Artforum, 18*(8 (April)), 58.
Fowkes, M., & Fowkes, R. (2022). *Art and climate change*. Thames & Hudson.
Glasheen, M. (1978). Uluru: Mythology of the dreamtime. *Cantrill's Filmnotes, 27–28*(Mar), 8–21.
Green, C. (2001). *The third hand. Collaboration in art from conceptualism to postmodernism*. UMP.
Green, C. (2008). The memory effect: Anachronism, time and motion. *Third Text, 22*(6), 681–697.
Heise, U. K. (2008). *Sense of place and sense of planet: The environmental imagination of the global*. OUP.
Kübler-Ross, E. (1973). *On death and dying*. Routledge.
Lipstadt, H. (2005). "Natural overlap": Charles and Ray Eames and the federal government. In D. Albrecht & B. Colomina (Eds.), *The work of Charles and Ray Eames: A legacy of invention* (pp. 150–177). Abrams Books.
Lorca Macchiavelli, C. (2019). *Vita Havet. Whiteness and otherness – Plaza de Mayo and Konstfack*. Masters thesis. University of Arts, Crafts & Design.
Mignolo, W. D. (2009). Coloniality: The darker side of modernity. In C. S. Breitwisser (Ed.), *Modernologies. Contemporary artists researching modernity and modernism* (pp. 39–49). MACBA.
Mirzoeff, N. (2014). Visualizing the Anthropocene. *Public Culture, 26*(2), 213–222.
Murray, R. L., & Heumann, J. K. (2009). *Ecology and popular film: Cinema on the edge*. SUNY Press.
Musser, C. (2014). Trauma, truth and the environmental documentary. In A. Narine (Ed.), *Eco-trauma cinema* (pp. 46–71). Routledge.
O'Hagan, S. (2017). Interview: John Akomfrah, Progress can cause profound suffering. *The Guardian*. https://t1p.de/dxi3n. Accessed 14 Nov 2023.
Richards, M. (2013). Greening wildlife documentary. In L. Lester & B. Hutchins (Eds.), *Environmental conflict and the media* (pp. 171–185). Peter Lang.
Serres, M. (1997). Science and the humanities: The case of Turner. *SubStance, 83*, 6–21.
Smith, T. (2009). *What is contemporary art?* UCP.
Smithson, R. (1968). A sedimentation of the mind: Earth projects. *Artforum* (September). Reprinted in Holt, N. (1979). *The writings of Robert Smithson* (pp. 100–113). NYU Press.
Warde, P., Robin, L., & Sörlin, S. (2021). *The environment: A history of the idea*. Johns Hopkins Press.

Chapter 9
Representing the Climate Crisis: Aesthetic Framings in Contemporary Performing and Visual Arts

Susanne Thurow, Helena Grehan, and Maurice Pagnucco

Abstract This chapter reviews the representation of climate change in performing and visual arts over the past ten years, canvassing the aesthetic exploration of the climate emergency in selected international works by surveying emergent narrative themes, key dramaturgical shifts and aesthetic strategies. Discussing the limitations of anthropocentric conventions, it investigates innovative approaches and their capability to generate knowledge about the dynamics of Earth processes and humanity's embeddedness and interference with them. Looking to novel experimental work currently in development at The University of New South Wales (UNSW)'s iCinema Research Centre, we speculate how these emergent aesthetics may be further developed to augment the arts's capability to deepen insight and strengthen preparedness in a rapidly transforming world.

Keywords Aesthetics · Anthropocentrism · Bruno Latour · Climate change · Contemporary visual arts · Performing arts · Preparedness · Timothy Morton

S. Thurow (✉)
iCinema Centre for Interactive Cinema Research, The University of New South Wales, Kensington, Australia
e-mail: s.thurow@unsw.edu.au

H. Grehan
Western Australian Academy of Performing Arts, Edith Cowan University, Mount Lawley, WA, Australia
e-mail: h.grehan@ecu.edu.au

M. Pagnucco
School of Computer Science & Engineering, The University of New South Wales, Kensington, NSW, Australia
e-mail: morri@unsw.edu.au

D. Del Favero et al. (eds.), *Climate Disaster Preparedness*, Arts, Research, Innovation and Society, https://doi.org/10.1007/978-3-031-56114-6_9

9.1 Introduction

Global climates are the envelope for life on Earth. Understanding their nature and dynamics is of paramount importance for humanity's survival into the future. They intimately enmesh with, yet also far exceed, our human sensory and intellectual grasp, articulating what philosopher Timothy Morton calls a "hyperobject"—an entity so "massively distributed in time and space relative to humans" (2013) that its entirety eludes critical faculties. While we can observe some "local manifestations" (ibid.), such as indicators of intensifying climate change, hyperobjects pose representational challenges that cannot be resolved by scientific enquiry alone. Their vast scale and temporal extension resist meaningful human engagement (Schneider & Nocke, 2014). Performing and visual arts can significantly augment understanding and preparedness for a turbulent future by offering spectators the opportunity to viscerally and sensorially experience (from the safe space of the performance space or gallery) the violent dynamics of escalating climate change. They can engage audiences by eliciting empathy, shame, wonder, despair or, indeed, horror—to allow them to envisage things they may know about in the abstract, or via the news media, anew. They create a productive space within which to envision, create and render often frightening future realities in ways that can open up discussion about how—as societies and communities—we might better prepare for these events. Research into risk communication has demonstrated that people who have experienced hazardous events are more likely to take action to mitigate against future harm and potential losses; that is, it strengthens their preparedness (Thieken et al., 2007). By foreshadowing possibilities of how extreme climate events could evolve, we can begin to develop a dialogue that shifts the current predominant focus from repair to preparation. Indeed, we argue that the arts have the capacity to imaginatively project—rather than merely reflect—the sociopolitical and cultural possibilities of a world in transformation, as well as to probe the ontological dimensions of our embedded planetary existence to make sense of the changes that are afoot. With their disciplinary strength residing in designing spaces that direct perception and augment sensory experience[1] to inspire new understandings and concepts of reality, the arts are perfectly placed to lead preparedness efforts for a future that we are yet to adequately envision.

In this chapter, we consider the ways in which key performance and visual art works have so far explored the idea and actuality of climate change through acts of performative storytelling. Our focus is not solely on specific weather events, nor is it on works that prioritise the written word as text. Rather, it is on prescient works of performance and visual arts—whether they are mainstream plays or avant-garde productions, interactive or traditional installations—that home in on and invite new considerations of the feedback loop that inseparably intertwines climate extremes and societies. In the process, we canvass dominant narrative themes, key dramaturgical shifts and aesthetic strategies employed to think through the dynamics of these conditions affecting human and non-human actors.

[1] That is, aesthetics

9.2 State of Play: Performing Climate Change and Planetary Interconnectedness

While a sustained and globally visible "ecological turn" has been slow to take hold in the performing arts (Chaudhuri, 1994), practitioners have often productively leveraged the trope of climate change to offer poetic commentary on the state of human social relations. For example, Andrew Bovell's deeply moving play *When the Rain Stops Falling* (2008) deploys shifts in global weather patterns as evocative backdrop on which the family drama unfolds. This amplifies and adds nuance to the narrative turns and allows the story to resonate into larger cosmological contexts. While some scholars (e.g. Ahmadi, 2015) read such work as prefiguring critical engagement with new materialist ontologies such as Morton's, the dominance of the human reference frame complicates such claims. The "climate"[2] here only emerges as an extended metaphor for the temporal condition of human life and sense making as it is reflected in the family story. Unpredictable weather in the play is metaphorically linked to the pedophilic nature of the family patriarch Henry Law, whose transgressive acts in the play's earliest time layer throw the fictive world off centre. The relentless rain only ceases once his story is fully uncovered and reckoned with by his grandchildren in the year 2039. In the play, the climate does not acquire its own agential force, and its representation ultimately reinscribes anthropocentric narrative traditions. These privilege the human as vortex from which ultimately all narrative focus, action and meaning spring—mirroring the dramatic traditions of the Global West, but also its scientific and philosophical histories with which the arts have co-evolved. The harbingers of catastrophic climate change, however, have been catalysing a comprehensive revision of anthropocentric worldviews, revealing the need to account more productively for the constitutive relations that weave the human and non-human worlds into an indistinguishable entity that far exceeds mere ecological coexistence (e.g. Bennett, 2010; Latour, 2018). With climate change announcing as unpredictable extreme weather across ever more locales around the globe, its temporal and spatial reach demand a reckoning that can no longer safely deflect to the "far away" or distant prospect of an only potentially hazardous future (Schneider & Nocke, 2014). It requires thinking beyond hierarchical metaphors and simplistic linear concepts of agency that render the climate an external force impacting the human world. A nuanced account is needed that acknowledges expansive impact yet also articulates the human as an out-of-control geophysical agent whose collective actions have altered the functioning of Earth systems. The age of the Anthropocene calls for a concept of agency that delineates the conditions of our planetary existence but also accounts for our impact as well as the limitations of our agency as part of a deeply interconnected world.

[2] In Morton's sense of "more or less a container in which objectified things float or stand" (2013)

9.2.1 Eulogising the Present

Developing and staging intricate ecological interconnectedness without reinscribing an anthropocentric ontology is a challenge that artists have been addressing in various ways—spanning a disparate continuum of narrative and dramaturgical experimentation. For example, Chantal Bilodeau's series of plays *The Arctic Cycle* (since 2014) has generated a layered panorama of the Arctic states, teasing out their complex fragile ecologies as places already living through the planetary future that global warming is increasingly visiting on a grander scale. Aspiring to co-create the plays with local community leaders, activists, politicians and scientists, Bilodeau weaves multimodal tapestries of spoken word, music, dance and puppetry that capture the intersectional nature of challenges flowing from the receding ice as well as attendant economic and cultural upheavals. As a non-Indigenous practitioner, she thereby interlaces her plays with thematic and dramaturgical references to Indigenous ontologies, incorporating aesthetic anchors (e.g. Inuit sound architectures) that resist a straightforward anthropocentric reading—instead imbuing the frozen landscapes with vitality and transformative agency (Balestrini, 2020). This sites the human story within an independent, animated cosmos that forestalls reductive readings and instead draws attention to overlooked dependencies. In the context of the Global South, Taofia Pelesasa's *Te Molimau* (2019) adopts a similar approach, transporting the presumably "far away" implications of climate change into the geographical and temporal "right here". Staging the final farewell from her Pacific Country of Tokelau in the year 2060 as the sea casts its blanket over the beloved home, protagonist Vitolina performs a last fatele to honour and reconnect with her culture that—in the very act of staging—reaffirms its enduring strength and orientational capability in the theatrical as well as the audience's felt present. Paying homage to all the world stands to lose in the face of hesitant (non)action on global warming, Pelesasa draws her audience into the fold of an interconnected, life-affirming ecology in which humanity finds an integrated home among a sprawling web of organic and inorganic matter. Framed as a eulogy, the play calls the audience to action in the present to prevent theatrical fiction from turning into a bitter future lived reality.

Marrugeku's *Cut the Sky—Five Songs for the Future* (2015) is an epic-scale performance work that equally "dances forward" the implications of our present actions. It explores strategies for ontological reorientation based on the concept of "Bugarrigarra [creation time that is ever present]",[3] which Dalisa Pigram, Rachael Swain and their team use as dramaturgical framework for their work (Pigram & Swain, 2021). Rendered in non-chronological vignettes that traverse the past, present and future, the work's unifying centre is provided by evocations of the Kimberley landscape of northwest Australia, within which it was created. Trans-Indigenous

[3] The Yawuru people define Bugarrigarra as "the time before time, when creative forces shaped and gave meaning to the landscape, putting the languages in the country for the people and creating laws" (Nyamba Buru Yawuru, n.d.).

storytelling[4] (i.e. dance, poetry, song and large-scale video projection) induces a deep desire for audiences to consider the vulnerability of the land as a result of human intervention and makes a plea for a different future. As creators Dalisa Pigram and Rachael Swain note, the intercultural work "is a meditation on humanity's frailty in the face of our own actions. In a burnt landscape, a group of climate change refugees face yet another extreme weather event" (2021). The focus is on the Earth's surface, the sky and what lies below—with specific consideration of the impact of mining in the Kimberley at the core of the storytelling. *Cut The Sky* moves forward and back in time in an "attempt to come to grips with the sprawling nature of climate change—and with who or what is in control as we ask how we value what is above and below the earth's crust" (ibid.). Its fragmentary, nonlinear and relational aesthetic that subtly explores "custodianship [and] mythopoesis" (White, 2015) thereby implies that answers to this question can never be found in definite, exhaustive or one-sided terms but must account for the multivariate web of relations that exist between the human and non-human across interlacing temporal and spatial scales.

9.2.2 *Performative Aesthetics*

David Finnigan's *Scenes from the Climate Era* (2023) is a play that comprises 50 short scenes performed within an 80-minute timeframe. In its world premiere at Sydney's Belvoir Street Theatre, Carissa Licciardello's staging embraced an "aesthetics of scarcity", comprising five actors on a bare stage without any props other than simple chairs and a table. It anticipated and realised at small scale the play's key act of advocacy, which is to reduce humanity's global carbon footprint (Flack, 2023). The vignettes deal with myriad topics, including a family attempting to live through an extreme heatwave of daily 55 °C with only a single, failing air-conditioning unit, or characters seeking to escape a flash flood in a small coastal town. The tone is conversational, humorous at times, as ethical and practical questions arising from each decision are weighed up—resonating with international productions such as Dawn King's *The Trials* (2022). The play makes clear that—while the world itself will never end—the future of humanity and all life forms will be bleak if we do not take tangible and decisive action on all fronts. The inclusion of debate and questioning creates a kind of bargaining with the audience to catalyse meaningful action in the present, before it is too late. As characters' reasoning and reference points flounder in the face of extreme events, they mirror the inadequacies of our contemporary mindsets—ill-equipped as we are to productively prepare for the world humanity has been shaping over the past centuries. As Eamon Flack argues, *Scenes* develops a blueprint for a climate change dramaturgy—one that abandons Aristotle's unity of time and place as well as focus on an individual's

[4] Bridging Yawuru, Nyikina, Walmajarri and Bunuba storytelling traditions and ontologies

Fig. 9.1 *Scenes from the Climate Era*, feat. Ariadne Sgouros, Harriet Gordon-Anderson, Charles Wu, Brandon McClelland and Abbie-Lee Lewis, image by Brett Boardman.

decision making, in favour of a sprawling, disjunct and collective portrait of crumbling Earth systems that dislodge humanity from any solid reference frame into a fluid cosmos of extended transformation (2023). By foretelling likely scenarios of the future, Finnigan probes what societal and cultural changes await beyond our present horizon and develops a dramaturgical framework that may help us navigate these violently turbulent times. This work engenders points of contemplation that invite audiences to see and feel the reality of climate change anew, from different vantage points, calling for conscious engagement with our present predicaments (Fig. 9.1).

Some artists integrate direct intervention modalities into their work to catalyse immediate action, adopting a politically engaged aesthetic. For example, Uta Kögelsberger developed the layered multimodal project *Fire Complex* (since 2020) on US wildfires, which comprises myriad artistic elements, each in its own way arresting. One of its publicly most perceptible outputs has been the video work *Cull* that documents the felling of ancient sequoia and conifer trees that were destroyed in the Californian wildfires of 2020. It was shown as a five-screen installation work in galleries; yet most notably, it was realised on giant billboards in downtown Los Angeles. As Kögelsberger notes, it directly became "an agent in the public realm. Set in stark contrast to the urban environment, where the impact of climate change often seems a remote abstract reality, [it offered] a timely reminder of the impact of our daily actions". Some of the filmed trees were up to 25 stories tall and over 2000 years old. The cinematic-scale, quiet and powerful images of these giants of the forest toppling over are almost balletic in style, yet in its eulogising tone also devastating. As Hettie Judah notes, the experience of watching *Cull* is "like a disaster movie. [...] As the trees fall, they smash through the branches of their smaller neighbours, sending out shockwaves as they hit the ground" (2022). Kögelsberger combines activism with art and has used the *Fire Complex* project to raise funds for sustainable reforestation, having helped plant over 6000 trees since the project began. This approach reflects her sense of dismay that not enough is done in preparation for extreme fire events with key focus still on recovery rather than regeneration and long-term sustainable repair. Kögelsberger's work has helped raise awareness of and funds for regeneration initiatives and garner broad public support that has led to the adjustment of US environmental laws aimed at reducing the ferocity of future blazes (Kögelsberger, n.d.). While this is an important step forward, destructive wildfires will nevertheless continue into the foreseeable future, despite reforestation and more successful forest management. Hence, we need Kögelsberger's activist approach as well as creative methodologies that can immerse us in extreme events to teach us how to prepare for and navigate such intense experiences.

A multipronged project that sought to address the urgent need for a shift from a culture of repair to one of preparedness was *Refuge* (2016–2021), organised by Arts House Melbourne. As discussed by Davidson et al. in this volume, it invited people affected by or working to mitigate the impacts of climate change to bring their knowledge to bear on creatively imagining life worlds in the immediate future. Taking stock of present changes in the environment, artists like Latai Taumoepeau and Keg de Souza extrapolated from these palpable yet often overlooked trajectories. They generated performative scenarios: for example, staging drill trainings such as converting North Melbourne Town Hall into a temporary flood emergency relief centre or inviting community to contemplate scenarios of food and water shortages in compounding crises (Arts House, 2021). These stagings derived their plausibility and affective power from persuasive presentation of evidence that the "here and now" already contains the seeds of future devastation. Looking closely and listening attentively became the key to documenting change over time, while performative aesthetics[5] allowed translating this data into meaningful scenarios that convey the future as an immediately felt reality. They gave possible futures visible shape and sensorial potency in the present, which opened people up to acknowledging dimensions of reality that they have not been perceiving yet—or even might never be capable of observing first-hand due to the limitations of the human sensorial apparatus.

[5] That is, structures that encode dynamic relationships between human or non-human agents in a work

9.2.3 Scaling Space and Time

Visual artists such as Jakob Kudsk Steensen, ScanLAB and Richard Mosse have dedicated much of their creative practices to expanding the limits of human perception and experience, investigating microscopic and durational methods to foster deeper appreciation for the interconnected nature of planetary life, its beauty and fragility. Novel experimental imagining and sonification techniques thereby provide the key to unlocking dimensions inaccessible to the naked eye—augmenting everyday experiences and prototyping new forms of engagement with the more-than-human world. In his large-scale installation *Broken Spectre* (2022), Mosse arranged 20m wide-screen panoramas that provided an arresting visual spectacle of the Amazonian rainforest and the transformations wrought by human interference. The entrancing ultra-high-resolution imagery (ranging from intense close-ups of forest floor to sweeping panoramic aerial shots) was in part generated by custom-made multispectral cameras that capture wavelengths outside of the range of human vision. Mosse and his team translated their information into fluorescent colours that add new narrative dimensions to the available data, deepening insight into the invisible transformations afoot in the complex tropical ecosystems. The result has been a dense semantic and politically engaged layering, which elucidated the complex interdependencies that cut across scales, from socioeconomic to biological, from micro to macro. It approximates a solution to the problem of representing climate change as a hyperobject that unfolds at a "scale [...] that is too vast to comprehend, too minute to perceive, and too normalised to see" (Mosse qtd. in Blue, 2022).

Similarly, ScanLAB's syncopated screen installation *Framerate: Pulse of the Earth* (2022) utilised a LiDAR 3D scanning system to document British locales over two years in millimetre-precise 3D time-lapse videos—revealing change hidden in time that we as humans are ill-equipped to perceive, let alone grasp. Technological innovation here allowed augmenting the means by which we can try to make sense of the climate emergency, complementing our direct sensory impressions with a remotely derived vision that connects the vast and minute scales of climate change in ways that are intuitive and compelling.

Jakob Kudsk Steensen's immersive live-simulated environment *Berl-Berl* (since 2021) lifted such experimentation to another level, using photogrammetry to generate a 3D-modelled synthesised macro-rendition of the wetlands surrounding the German capital, Berlin—blending present-day scans with reconstructions of species that have since vanished. While the camera relentlessly auto-pans over the simulated landscape, it activates sound layers mapped to features in the environment. These are, in turn, embedded into larger algorithmic structures that reflect the soundscapes of broader dynamic systems, like particular weather conditions and the circadian rhythm of the natural world. The synthesised sound is played through a network of 29 speakers dotted across the gallery space (Chapter TV, 2021). As all animation is live-simulated, each installation experience is unique and has its own unpredictable tone and affective resonance. Fostering emotional states is conducive to eliciting contemplation and reflection in the viewer—a prerequisite for any shift in consciousness and behaviour to occur. Aesthetic experimentation here translates invisible processes into perceptible entities and transforms our sense of reality by establishing new relations that become available to structure our imagination and pathways for preparatory action.

Highlighting the co-constitution of ecosystems and exploring their self-organising mechanisms is another potent way to invite audiences into an ontological reconsideration of interconnected planetary ecologies. A stand-out example in this regard is Pierre Huyghe's *After ALife Ahead* (2017). The work consists of a complex arrangement of carefully engineered biological and technological systems, which are brought together in such ways that state changes in one system trigger marked transformations across all others. The work was staged in a disused ice rink in Germany and consisted of earth mounds, aquariums, live animals,[6] cancer cells in incubators and a range of sensors that—based on data provided by the live organisms—trigger movement of physical and virtual portals in the space. The chain reactions highlight the plasticity and self-organising capabilities of evolving ecosystems, yielding a dynamic tableau of perpetual becoming that, albeit conceived and installed by the artist, henceforth eschews his directive control. The interdependency between systems, catalysed by exposure to the elements (through a permeable ceiling that admits rain and sunshine), means that constant adaptation takes place. Technology (in the form of mechanical and digital architectures), rather than designed to control the biological and atmospheric processes, is horizontally integrated—with AI algorithms scripted in such a way that they shift in response to certain state changes in the space. The AI seamlessly integrates with biological systems—bridging organic and inorganic matter, which effectively muddies any distinction between natural and cultural systems (Frohne, 2020). The result is a symbiotic chaos that discloses its interdependency and responsiveness, yet in its complexity and unpredictability must remain enigmatic to the human epistemological quest. Thus, Huyghe's work can be read as a critical commentary on the anthropocentric assumption of human primacy and our quest for control, embedded in knowledge traditions that permeate much of Western art history.

An anthropocentric perspective blinds us to the interconnected nature of those agencies that determine the boundary parameters of our planetary existence. Accordingly, Bruno Latour identified the need for a "new climatic regime" that pays attention to the interrelated Earth systems that condition our existence—i.e. atmosphere, biosphere, hydrosphere and lithosphere (2018)—with "Climate" denoting the output of their dynamic interactions. Arguing that these interrelated processes (and by extension, the Climate) have been fermenting under intensifying global warming, he proposes the concept of "terrestrial agency" to capture the dynamics that underpin the ferocious floods, earthquakes and wildfires that are arriving as compounding disasters across the globe (2018). As a constituting force, the terrestrial intervenes into the fabric of our lifeworlds, forcing an ontological repositioning that acknowledges humanity's circumscribed role in the wider planetary ecology.

[6] Including chimaera peacocks, snails, algae, bacteria, bees, etc.

Even though the cumulative impact of our extractive practices has altered the long-term functioning of Earth systems, our ability to intervene into these ad hoc to remedy the resulting fractures remains at best limited and subject to a techno-optimistic fallacy. To work against this "slow cancellation of [our species'] future" (Fisher qtd. in Frohne, 2023), we need to cultivate a critical participatory relationship with the terrestrial, learning to perceive, relate and interact with its agency. In the performing arts, Pierre Daubigny explored the possibility of engaging with such a long-term ontological shift through an expanded emotional repertoire in his experimental play *Gaïa Global Circus* (2013).[7] The work seeks "to capture, understand, feel, and represent to oneself the irruption of the new [Terrestrial] character" (Aït-Touati & Latour, 2018) and to utilise the full range of our knowledge-producing modes, i.e., "our bodies, our senses, our tools, our instruments, and all that allows us more generally to capture and assimilate the world" (ibid.). Daubigny and his team engaged in a collective writing process that involved scientists, activists and artists. They drew inspiration from scientific practices of "sampl[ing], collect[ing], measur[ing], analys[ing], transform[ing] and styl[ing]" material into the stage work. Fostering intense dialogue behind the scenes as well as on stage between stakeholders, this practice allowed practical, intellectual and affective modes to coexist and cross-pollinate. The tone of the "kaleidoscopic text" (ibid.) eschews "[didactic] discourse" and eulogising "*lamento*" to foster a "burlesque imaginary" that absorbs the often paralysing political posturing of climate change art through a rapid turnover of pluri-vocal vignettes. The frenetic stage action thereby competes with, and is ultimately dwarfed by, a tone-setting set design that consists of a canopy that envelopes both stage and auditorium. Sensitive to disturbances in the space (e.g. rising temperatures due to collective body heat), the eye-catching white sheet performs its own mesmerising dance that is enmeshed with yet also markedly set apart from the human drama.

9.3 Novel Experimental Trajectories

Our review demonstrates that artists have sought to render palpable the implications of planetary degradation through manifold creative approaches. While testament to a growing awareness that the climate emergency will define our species' ultimate survival into the future, much work remains conceptually and aesthetically underdeveloped. Most artists tend to position audiences as passive onlookers faced with climate change scenarios that are directly extrapolated from present trajectories. Very few works immerse audiences in the actual dynamic, multisensory unfolding of events, let alone speculate on what these might evolve into as the world is heading towards a 2.9 °C warming by the end of this century (United Nations Environment Programme, 2023). The unpredictable interaction of Earth systems under such

[7] Co-developed with Bruno and Chloé Latour, as well as Frédérique Aït-Touati

conditions takes current scientific modelling capabilities to their outer limits, requiring novel aesthetic approaches that may map the shape of this future world (Shepherd et al., 2018). We argue that the arts can provide a productive space to envision such possibilities, to furnish evocative scenarios that allow visualising this world that awaits us at the threshold between "now" and this yet un-envisioned "then". Such creative practice has to "reach[–] beyond the hermeneutical to the ontological" (Elias & Moraru, 2015) and generate poetic sensory experiences of planetary interconnectedness that anchor knowledge through embodied rather than solely cognitive sense making.

An example of emerging research that seeks to address this challenge is the large-scale *iFire* project (2021–2025), led by Dennis Del Favero at UNSW's iCinema Research Centre. The project has three distinct modalities: artistic, scientific and industrial.[8] Each modality leverages machine learning (ML) and artificial intelligence (AI) to analyse existing wildfire data. The aim is to develop a generative AI system that can aesthetically render and plausibly upscale on screen the increasingly erratic and unpredictable behaviour of real-life wildfire events. The project's key undertaking is to translate these dynamics into high-fidelity 3D landscapes that are digitally twinning real-life places, which viewers can interactively curate and explore in real time across a range of scalable platforms—including laptops and 360-degree CAVE theatres. In the latter, users find themselves "virtually surrounded by a 3D-[reconstructed] forest landscape, [becoming] close witness of an emergent forest fire whose powerful and rapid expansion [they] cannot escape" (Frohne, 2023). The visceral effect of facing the unforeseen behaviour of "flames, flying sparks and the suction of the draught created by the raging" blaze is enhanced by a dramatic soundscape of roaring fire and searing leaves, seeking to provoke "an existential moment, an immersive experience of anxiety" (ibid.). Here, users are placed at the heart of an extreme event of which no direct experience would be possible in the physical world—letting them viscerally explore, and so prepare for, its escalating dynamics from within the safe simulation environment. In future iterations, viewers will be able to adjust variables, such as temperature, humidity, wind speed or direction, and observe the change in fire behaviour—flames may travel faster and reach higher, spread more erratically and be carried by ember storms, racing through the canopy, consuming vegetation layers and leaving smouldering moonscapes. All the while, the AI system will analyse viewer decision making and situational awareness and respond to their interaction by enhancing or challenging it, imbuing scenarios with open-endedness.

For the artistic modality, Del Favero drew on *iFire*'s visualisation methods to develop the art installation *Penumbra* (2023), which experimented with a palette of aesthetic intensification. It adopted a monochromatic black-and-white aesthetic, in which the moving imagery was accompanied by an emotionally heightened voiceover of extracts from Georg Büchner's *Lenz* (1836), a prophetic fictive account

[8] That is, exploring a new artistic paradigm of intelligent interaction, provisioning an augmented visualisation platform for fire behaviour analysis as well as furnishing a training platform for firefighters

Fig. 9.2 Exhibition view of *Penumbra, iCinema (2023)*

of the encounter between human and climatic forces. Like *Lenz*, *Penumbra* is set in the Vosges mountains. Its voiceover conveys a "breathless account of a self-dissolution" in which the narrator spirals into "a state of insanity" (Frohne, 2023). The human suffering here allegorically intertwines with a forest's devastation, articulating the coupling of human and terrestrial agencies into which the viewer enmeshes through their motion-tracked interaction. Contrary to Bovell's representation, the simulated forest here exists as an independent entity with its own embodied and goal-orientated trajectory. The wildfire enacted in *Penumbra* "redefine[s] nature as a site of aesthetic-conceptual speculation" (ibid.), with immersive interactive visualisation used as a tool to generate a contemporary panorama of the world's ontology of which the human is only but one agent (Fig. 9.2).

9.4 Conclusion

Penumbra probes the aesthetic capabilities of immersive interactive visualisation to transform the climate crisis from the detached metrics of science into tangibly personal experiences that seek to enhance meaningful engagement with the cascading impacts of global warming. Its underlying data repository as well as conversion, visualisation and programming methods will be used to inform new research into poetically rendering plausible future scenarios with the aim of augmenting our preparedness in the face of violent climatic transformations. Instead of treating climate change as an ambient weather phenomenon that acts as a backdrop or news arena, such new research will continue the lines of enquiry pursued by artists such as Mosse and Huyghe into novel visualisation and spatial aesthetics that can immerse viewers in dynamic terrestrial transformations. Drawing inspiration from the dramaturgical and narrative strategies emerging in prescient dramatic work such as *Scenes from the Climate Era* and *Refuge*, this research will aim to probe the performative dimensions of these transformational processes. It will explore modes of embodied enactments that can aesthetically capture and convey the interconnected ontological condition of planetary existence—prototyping visceral encounters that meaningfully engage the human as a terrestrial agent. For example, it may investigate the performative potency of heat as a catalyst for dissolving and reshaping connections—building on *iFire*'s wildfire aesthetic to reveal the underlying processes of pyrolysis and how these may be dialogically articulated through viewer presence and interaction through AI and ML in ways resonant with Daubigny's approach.

The broader aim of this comprehensive research programme would be to analyse emerging means for representing and engaging with the hyperobject "Climate", to explore the implications and meaning of their respective aesthetics and to probe what new relations may become available to guide imaginative preparedness as a result of such enquiry. Performing and visual arts, as well as creative technology, are here enlisted as drivers of speculative practice that connects the threads of familiar environments with the possibilities emerging from the latest climate modelling to anticipate and respond to a future we are as yet unable to envisage. It expands the concept of an "ecological turn" beyond thematic and dramaturgical concerns, reimagining how the arts as a disciplinary ensemble may augment readiness for the unchartered volatile conditions that humanity will face on this heating planet in which we are embedded.

Acknowledgements Dr. Susanne Thurow's Laureate Postdoctoral Fellowship is funded by the Australian Research Council Laureate project FL200100004 (directed by Laureate Professor Dennis Del Favero), which develops the *iFire* research. Professor Maurice Pagnucco is an expert collaborator on *iFire*.

References

Ahmadi, M. (2015). Andrew Bovell's 'When the rain stops falling': Theatre in the age of 'hyperobjects'. *Australasian Drama Studies, 66*, 40–62.

Aït-Touati, F., & Latour, B. (2018). Performing Gaia. *Theater, 48*(2), 95–101.

Arts House. (2021). *Refuge*. ArtsHouse.com. https://t1p.de/tdzbj. Accessed 20 Nov 2023.

Balestrini, N. (2020). Sounding the Arctic in Chantal Bilodeau's climate change plays. *Nordic Theatre Studies, 32*(1), 66–81.

Bennett, J. (2010). *Vibrant matter. A political ecology of things*. DUP.

Blue, A. (2022). The NGV presents Richard Mosse: Broken Spectre. *To Be Magazine*. https://t1p.de/4xylh. Accessed 20 Nov 2023.

Chapter TV. (2021). *Introducing Jakob Kudsk Steensen: Berl-Berl live simulation*. Youtube.com. https://t1p.de/nonur. Accessed 20 Nov 2023.
Chaudhuri, U. (1994). "There must be a lot of fish in that lake": Toward an ecological theater. *Theater, 25*(1), 23–31.
Elias, A., & Moraru, C. (2015). *The planetary turn: Relationality and geoaesthetics in the twenty-first century*. Northwestern UP.
Flack, E. (2023). Foreword. In D. Finnigan (Ed.), *Scenes from the Climate Era* (pp. 2–4). Australian Plays Transform.
Frohne, U. (2020). After a life ahead—Inszenierung einer Welt nach uns in. In L. Crasemann, B. Fellmann, & Y. Hadjinicolaou (Eds.), *Seismographen und Orientierungsspiegel* (pp. 306–312). DeGruyter.
Frohne, U. (2023). Night drifts above the Earth. In I. P. di Persano (Ed.), *Dennis Del Favero. Viscera* (pp. 28–31). Recontemporary.
Judah, H. (2022). 'Like a disaster movie'—the film about the felling of California's giant fire-poisoned sequoias. *The Guardian*. https://t1p.de/qxptc. Accessed 6 Nov 2023.
Kögelsberger, U. (n.d.). Fire Complex works. *utakoegelsberger.UK*. https://t1p.de/exh7g. Accessed 6 Nov 2023.
Latour, B. (2018). *Down to earth. Politics in the new climatic regime*. Polity.
Morton, T. (2013). *The end of the world. Hyperobjects: Philosophy and ecology after the end of the world*. UMP.
Nyamba Buru Yawuru. (n.d.). Mabu Liyan. Good Spirit. Yawuru.org.au. https://t1p.de/43186. Accessed 15 Dec 2023.
Pigram, D., & Swain, R. (2021). Cut the sky. In H. Gilbert, D. Pigram, & R. Swain (Eds.), *Marrugeku: Telling that story: 25 years of trans-Indigenous and intercultural performance* (pp. 254–279). Performance Research Books.
Schneider, B., & Nocke, T. (2014). *Image politics of climate change: Visualizations, imaginations, documentations*. transcript.
Shepherd, T., Boyd, E., Calel, R., Chapman, S. C., et al. (2018). Storylines: an alternative approach to representing uncertainty in physical aspects of climate change. *Climatic Change, 151*, 555–571.
Thieken, A., Kreibich, H., Müller, M., & Merz, B. (2007). Coping with floods: preparedness, response and recovery of flood-affected residents in Germany in 2002. *Hydrological Sciences Journal, 52*, 1016–1037.
United Nations Environment Programme. (2023). *Emissions gap report 2023*. UN Environment Programme.
White, T.-A. (2015). Cut the Sky. *Australian Book* Review. https://t1p.de/4p2de. Accessed 3 Nov 2023.

Part III
Rehearsing

Chapter 10
Supporting Disaster Preparedness Through User-Centred Interaction Design in Immersive Environments

Alethea Blackler, Nagida Helsby-Clark, Michael J. Ostwald, and Marcus Foth

Abstract At a time when wildfires and severe floods are challenging human society in unprecedented ways, we examine how immersive virtual environments can be used to enhance community preparedness for, and engagement with, disaster scenarios. Drawing on research from the fields of interaction design and participatory design, we explore the capacity of three-dimensional (3D) immersive virtual environments to foster increased situational awareness and risk perception among diverse communities—from first responders to local populations. Investigating tangible interfaces and interaction schemas applied to spatialised settings, we demonstrate how immersive environments can support effective scenario testing and rehearsal of responses to hazardous situations. Application of the described methods can equip users with response strategies that may prove productive in augmenting risk perception and deliberation.

Keywords Digital inclusion · Disaster preparedness · Immersive design · Intuitive interaction design · Participatory design · Scenario development · Serious games · User engagement

---

A. Blackler
School of Design, Queensland University of Technology, Brisbane, QLD, Australia
e-mail: a.blackler@qut.edu.au

N. Helsby-Clark
iCinema Research Centre, The University of New South Wales, Kensington, NSW, Australia
e-mail: n.helsbyclark@student.unsw.edu.au

M. J. Ostwald
School of Built Environment, The University of New South Wales, Kensington, NSW, Australia
e-mail: m.ostwald@unsw.edu.au

M. Foth
School of Design and Digital Media Research Centre, Queensland University of Technology, Kelvin Grove, Australia
e-mail: m.foth@qut.edu.au

D. Del Favero et al. (eds.), *Climate Disaster Preparedness*, Arts, Research, Innovation and Society, https://doi.org/10.1007/978-3-031-56114-6_10

10.1 Introduction

In our contemporary world, climate disasters are having an increasingly catastrophic impact on communities. Yet, much focus of existing research and investment has been on "resilience"—broadly defined as the capacity to recover from an extreme event. While resilience is undoubtedly a crucial factor, of equal, if not greater, importance is "preparedness". In 1735, Benjamin Franklin famously warned the people of Philadelphia, who were facing an unprecedented threat of fire, that "an ounce of prevention is worth a pound of cure" (Franklin, 1735). Being able to foresee a problem allows us to better prepare for it and reduce its negative consequences. At a time when wildfires and severe floods—that were previously classified as "one in 50-year events"—now occur annually, preparedness is at least as important as resilience. However, while research and investment into disaster resilience have grown over the years—including in interaction design (Satchell & Foth, 2011)—this is not the case for preparedness, where few innovative methods have been developed to improve the ways both the wider community and first responders can plan for extreme events. This gap in research is a catalyst for us to examine how user interactions in immersive virtual environments can support innovative and effective strategies for disaster preparedness.

The chapter commences with an examination of the role of intuition and experience in preparedness. It notes how being able to visualise, describe and communicate the experience of an extreme scenario allows people to better rehearse how they will respond to the real event. The chapter examines how the role of vicarious experience, developed using immersive 3D experiences, can be effective for (i) visualising and enhancing preparedness and (ii) developing narratives or stories to engage users. We then consider the creation of scenarios to accommodate the visualising, describing and rehearsing of extreme event responses, along with the technology required to achieve this. The primary approaches described in this chapter use augmented reality (AR), virtual reality (VR) and serious games (SGs). These have the potential to foster increased situational awareness and risk perception within communities—from first responders to local populations. We conclude with a discussion of future research required to further improve immersive environments for disaster preparedness.

10.2 Intuition and Experience

Preparedness is a function of prior experience and of perceived threat (Lazo et al., 2015). While direct experience can be a strong motivator for enhancing one's preparedness, the majority of people will likely encounter major extreme events (e.g. a large earthquake) without such prior exposure. They will need to rely on experience of smaller events (e.g. earth tremors), different types of disasters (e.g. flash floods), other adverse life events (e.g. accidents) or vicarious experiences (e.g. media reports or accounts of prior events by people affected) (Becker et al., 2017). Research found that such experiences inform people's understandings and actions in relation to disasters (ibid.). Furthermore, the work of Klein (1998) and others into the experiences of first responders and disaster experts has shown that prior participation in similar events can activate intuition in a stressful situation that requires a time-critical response. It allows fast and generally accurate decision making (Miller, 2018). Professionals trained in such a way can draw on both "on-the-ground" experience and learned expertise to adapt to extreme situations (ibid.).

However, professionally trained first responders are no longer the dominant support force during extreme events. Fires and floods have become common in remote, peri-urban and suburban areas, where neighbours often attempt to help each other out of danger (Shaw et al., 2013). Members of the public often do not have relevant prior experience or expertise to draw on and often lack a rehearsed plan. They tend to receive only general instructions and advice in time-critical and stressful situations (e.g. through the media or emergency services broadcasts). Becker et al. (2017) found that the more direct peoples' experiences of a disaster were, the more likely they were to relate to it in active terms (i.e. display threat awareness, demonstrate actionable knowledge and engage in interpersonal discussion). Equally, people would show a heightened understanding of its consequences and use future disasters as a frame of reference, form adequate beliefs and emotional connection and have a stronger motivation to prepare. Importantly, Becker et al. (2017) found that vicarious experience does not achieve the same level of personal hazard experience as direct or indirect experience of an event. As it would be unethical to subject people to extreme and dangerous events in order to equip them with first-hand experience, other means of fostering productive exposure are required. A way to provide a safe and powerful vicarious experience is to develop immersive interactive virtual environments that can realistically simulate an event—promising to facilitate a more visceral experience by making the participation feel as if they are present in the event—than, for example, merely reading a story or watching a film.

10.3 Vicarious Disaster Experience via VR and AR

Professional disaster response training commonly includes VR simulation training to facilitate vicarious experiences of various emergency situations (e.g. Hsu et al., 2013; Nakanishi et al., 2009). Such role-playing and similar exercises can develop the type of participatory involvement needed for building intuition that can support decision making in disaster situations (Miller, 2018). Hence, they have been an integral part of emergency services accreditation and training for many years (e.g. see Alexander, 2000). In domains such as teacher and medical training (e.g. Korucu-Kış, 2021, Stegmann et al. 2012), vicarious experiences have been shown to be as effective for developing the skills to respond to real patients or students as actual direct experience. For example, trainees may observe others undertake procedures on simulated patients, or they may reflect on critical incidents and dilemmas in online settings. Across the board, combining vicarious and "hands-on" experience was found to be the most effective approach (Stegmann et al. 2012).

Professional emergency response decision makers generally have extensive data at their disposal to support them in their practice. Yet, this data is often only available in impersonal numerical form, which complicates assessment and rapid conversion into practical situational understanding—limiting their ability to appropriately respond to time-sensitive incidents. Members of the public have access to less data and often no relevant experience that supports them in responding to the critical situations, so they have even less usable information in a crisis situation. For them, the chance to gain a vicarious experience of an extreme event could be lifesaving. However, offerings of immersive scenario-based interactive activities to the general public are extremely rare.[1] To facilitate vicarious experience for professional responders, volunteers and communities, a pool of relevant prior experience must be assembled, coded and presented so that people can draw on a library of possible actions when immersed in an extreme scenario—whether as residents, first responders or decision makers. For developing a VR training scenario, two methods are commonly used to present such experiences: scenario-based learning and serious games (SGs).

Alexander (2000) argues that scenario-based learning "is useful [for] developing skills [such as] time management, cognitive mapping, [mediation,] team management and decision making under stress". He defined a scenario in the disaster/emergency education context "as a reconstruction of past [events] or a hypothetical construction of future ones" (ibid.). When used as a teaching or training tool, scenarios are often employed to stimulate discussion or reflection (Korucu-Kış, 2021). Instructional methods using scenarios typically involve assigning only a small number of roles to participants. However, if the number of roles and their complexity are broadened, an emergency simulation game can be developed (Alexander, 2000). There is evidence to suggest that games can be even more effective than scenarios when training young professionals in disaster scenarios. Ma et al. (2021) used a randomised controlled trial to compare "the impact of theme games and scenario simulation on the disaster nursing competence of nursing students". They "found that the overall [scores for competence, cognition, skills and affective response] were significantly higher in the game-based teaching group than in the scenario-simulation teaching group" (ibid.). They attribute these findings to increased motivation and improved autonomy and feelings of competence for those in the game group.

A "serious game" is a game wherein the goal is knowledge development, typically to educate the "player", rather than merely to entertain. SGs allow for agency: the capacity to shape the results, rather than just role-playing a set scenario. The player in a SG is not just experiencing a situation or scenario and reflecting on it—they are acting in it, trying out different options for response. One example of a SG

[1] One exception is a training programme for children that teaches them how to respond to disaster events such as earthquakes (Hatakeyama et al., 2016).

for disaster preparedness is a game by Lovreglio et al. (2018). It uses VR and the gaming engine Unity to prepare people to evacuate a hospital in the event of an earthquake. The game environment was designed to allow interaction (with environmental features and non-player characters) and navigation through a real building (Auckland City Hospital). Lovreglio et al. (2018) note the complexity of simulating a natural disaster using just sound and vision, especially when movement (shaking of the ground), heat (from fire) or touch (water from sprinklers or flooding) is part of the actual experience. They also stress the importance of using "real world" settings and scenarios in SGs as a means of increasing engagement and learning. Tsai et al.'s (2020) research is part of a field known as "game-based disaster education", which has a focus on preparedness and delivery "at scale", meaning to large numbers of people in a community. They developed the game *Battle of Flooding Protection* to increase willingness and capacity to engage in disaster response activities. Tsai et al. (2020) argue that SGs are more engaging because they situate the participant in the scenario, allowing them to interact with and shape events alongside other people. SGs like these two examples typically award points or achievements to acknowledge successful outcomes or reward effective responses, activities or strategies. As such, SGs offer a means of engaging players in visceral immediate situations, potentially developing effective synthetic experience, which has the potential to fuel their intuition (Miller, 2018).

Not all scenario experiences need to be visually immersive to be effective. The examples presented thus far have been reliant on fully immersive VR experiences in labs or through head-up displays (HUDs). Yet, it is also possible to use AR (on phones with cardboard hoods and low-cost headphones) to create emergency scenario overlays of real communities. Such an approach may not have the level of technical finesse of the discussed SGs, but by using real-world locations that people are physically situated in, a different type of emotional engagement and interactivity can be achieved (Gonsalves et al., 2021). AR-enabled SGs prompt and facilitate the sharing of local knowledge among communities and provide a tool to rehearse localised disaster scenarios. Such techniques provide a useful co-design approach to rapidly iterating hyperlocal emergency response training. They may accompany and inform the design of more immersive narratives.

Tabletop AR sandboxes (see Fig. 10.1) are another approach to enabling more immersive encounters with scenarios such as intense floods. These devices use real sand as a tangible interface for participants to explore the movement of water through contours in a landscape (Sánchez et al., 2016). A colourful contour map is projected onto the sand, augmenting the physical sand with information to facilitate learning and exploration. Water is projected into deeper regions, mimicking the effects of fluid dynamics on the landscape. This approach allows participants to mould the sand into different heights and shapes, to observe the effect of terrain on shaping water movement and of rain in shaping terrain. AR sandbox installations invite easily accessible tactile participation and can support understandings of possible flood scenarios and geoscience more broadly. Sánchez et al. (2016) describe these experiences as being more intuitive and faster vehicles for community participation compared to computer simulations.

Fig. 10.1 AR sandbox used for modelling flood risk. City Analytics Lab, UNSW, Sydney (2022)

Whether we develop scenarios (with or without reflective activities) or games or some combination of the two, designers need to consider at least four types of challenges: (i) understanding stories and contexts, (ii) creating scenarios or games, (iii) working with technology to deliver them and (iv) navigating equity of access.

10.4 Understanding Stories and Contexts

A key consideration early on is how we garner or develop the scenarios so they are realistic enough to be both useful and engaging. Any design process must first seek to understand the nature and dynamics of the real-world scenario and its broader ecological, social, cultural, symbolic, spatial and temporal contexts. The designed scenario must be based upon the real-world conditions. We need to understand how people caught up in extreme events interact with the event, the available situational information, other affected people and life forms and surrounding artefacts (from their car to their phone, their hose or their house). Such information can be garnered through interviews with survivors and first responders; studying reports, books and media; and reviewing Google Maps or similar platforms. A useful skill is what Lertzman (2019) calls "radical listening", which combines empathy, open-mindedness, presence and validation to conduct a deep and transformative form of listening that goes beyond the surface level.

10.5 Creating Scenarios or Games

Creative approaches to prototyping as well as community engagement throughout the design process have proven to be key to designing successful immersive experiences that support preparedness. Communities affected by disasters consist of different types of stakeholders with different levels of technical literacy and access to equipment. These range from professional disaster responders, State Emergency Service (SES) and fire brigade volunteers to local populations of all ages in various settings, from city centres to remote communities. Understanding who users are and designing for them is vitally important—especially since interaction needs to feel natural and intuitive, while the scenarios estrange familiar territory and depict novel and uncomfortable situations.

Participatory design approaches recognise different users as collaborators with their own context and domain expertise and actively involve them in the design process. Participatory design denotes a set of methodologies that reimagine traditional boundaries between designer and user. While precise definitions remain contested, researchers agree that a core component of participatory design is a process of collaboration in which mutual learning takes place (Greenbaum, 1991; Robertson & Simonsen, 2013). Co-design is typically viewed as a practice within the wider umbrella of participatory design (Sanders & Stappers, 2008). From this perspective, the design process becomes a space for mutual learning between collaborators, software engineers and decision makers, facilitated by the designer. The latter deploy their own expertise in facilitating experiences, drawing on a broad toolkit of methods (Kerr et al., 2023). Applying this approach to disaster preparedness can enrich the design process, incorporating local and tacit knowledge that the stakeholders have of potential disaster contexts. Active involvement is also proven to engender more widespread community support, agency and ownership and increases the likelihood of adoption of design solutions (Blomberg & Karasti, 2013). Dialogue in the design process is crucial when intervening in complex local decision making and environmental contexts for seeking to enhance disaster preparedness. SGs can be effectively integrated into a co-design process, where they can prompt idea generation and sharing of experiences and local knowledge.

One approach to better capture community insights is to engage participants in workshops creating tangible mock-ups. These are typically low-fidelity and tangible visualisations of contexts, leveraging readily available materials. Ehn and Kyng (1991) emphasise that the role of mock-ups lies not in their proximity to the real thing but in their ability to support communication, reflection and discussion between participants. This allows community members to take part in the making process and to communicate their knowledge at speed, without needing to understand complex technology nor to rely on software engineers to code their ideas. Mock-ups can be vehicles for mutual learning and communication. For Sanders (2002), the expertise of users (as experts of their own experience) emerges in what they "do", "say" and "make". While traditional user interviews may convey observable and audible knowledge, prototyping can be a tool to access tacit user knowledge.

10.6 Working with Technology

Immersive experiences typically involve a spectrum of technologies, which Milgram and Kishino (1994) place on a "reality–virtuality continuum". On this continuum, AR overlays real environments with digital information, mixed reality (MR) blends the digital and the real-world environment, and VR immerses users in a fully synthetic environment where the real world is no longer visible. Collectively, these approaches are often called extended reality (XR). Suh and Prophet (2018) add to this list non-immersive virtual reality, which captures users accessing VR experiences from a desktop computer with a two-dimensional interface. The spectrum does not necessarily entail a greater degree of immersion at either end. Rather, Suh and Prophet emphasise that different technologies offer different qualities of immersion. The appropriate technology should be selected based on a range of factors, including the affordances of the modality and how well-suited they are to the user's specific needs and contexts.

Immersive technologies offer the potential to support preparedness rehearsal, simulating disaster events to the extent that users gain vicarious experience. For Ryan (2015) and others, immersion is not simply a property of the technology, but a perceptual experience engendered through interactivity and imagination. She posits immersion as an active process, which may be temporal, spatial or emotional, reflecting the importance of interactivity as a gateway to immersion. For Ryan, narrative immersion may also be evoked through a variety of texts, including books. A key goal and measure of immersive experiences is their ability to evoke a sense of presence. For Zhao (2003), the sense of "being there" (i.e. presence) is dependent on the users' ability to suspend disbelief. Arguably, the ability of these immersive experiences to transport users is more closely tied to the interactive narrative potential of the experience than the degree of photorealism (IJsselsteijn & Riva, 2003; Seegert, 2009). Indeed, Seegert (2009) interprets presence as performance. Designers may support users to construct interactive narratives in which they are viscerally transported, as though they are truly present in disaster situations. While an experience does not necessarily need to be photorealistic to support feelings of presence, the experience should be credible in order to not break the imaginative spell, which sees the user suspend disbelief (Seegert, 2009). This may depend on the particular user. For example, firefighters confronting a fire scenario should be able to trust that the scenario they are envisaging is realistic and accurately depicted. While immersive technologies are not the only way to engender a sense of narrative transportation, they present opportunities to more fully envelop a user's senses, as well as the opportunity to support highly interactive experiences through real-time game engines. Technologies such as XR and 360-degree cinematic environments may provide the opportunity to gain vicarious experiences of specific disaster scenario that would be too dangerous or difficult to replicate in the physical world.

Increasingly, experiences are not limited to one type of interface. New visualisation systems such as *iFire* (discussed by Del Favero et al in Ch. 2 of this volume) are examples of cross-reality (Maurer et al., 2022) immersive experiences designed to

Fig. 10.2 *iFire* visualisation system. iCinema Research Centre, UNSW, Sydney (2022)

support firefighters and laypersons to rehearse encounters with extreme wildfires (see Fig. 10.2). *iFire* is powered by Unreal, a real-time game engine commonly used in virtual production. This technology enables experiences to be adapted and deployed across a range of visualisation platforms, including projection-based 360-degree and 130-degree 3D cinemas through to desktop screens and tablets—with capability to also run on head-mounted VR displays. When designing for such an ecosystem of modalities, designers must consider different contexts of use. For example, Del Favero et al. (2022) emphasise the affordances of room-scale spaces (such as 360-degree cinemas) for providing social perspectives, which inform and mediate the experience of all participants, as they interact not only with the virtual environments but each other as well. Similarly, immersive tabletop experiences may encourage this sort of real-world collaboration (Ens et al., 2021). Head-mounted displays may provide a more portable experience—while they retain immersion, they usually only do so on an individual basis. Laptops support a more accessible and affordable, if constrained, immersive experience yet still facilitate a sense of immersion by allowing the user to feel present through their navigational control inside a 3D environment (Del Favero et al. 2023).

Engendering a comfortable user experience with what may be quite unfamiliar technology to users is a key challenge. *iFire*'s user groups are not necessarily passionate gamers or those familiar with VR. Within immersive cinematic environments, users interact with *iFire* via a tablet. Users can navigate through space and time to visualise environmental variables, such as wind, temperature and humidity, and to gain an understanding of their location in the virtual world. In future iterations, users will be able to adjust variables of the wildfire to generate plausible novel scenarios using the application's inbuilt AI system. The advantage of the tablet as a more tangible user interface is that, compared to a gaming controller, a wider range of users are already familiar with the interface designs, since they are widely used in professional as well as private settings.

Research into intuitive interaction with interfaces and products over several decades has shown empirically that the ability to transfer prior experience to a new situation is the main factor that determines how fast and accurate a user adapts to a new interface or object (Blackler et al., 2010, Fischer et al., 2015, Still & Still 2019). For example, the use of touchscreens on smartphones or tablets may support a more intuitive user experience. By incorporating familiar and accessible components, these interfaces may support communities to more readily adapt to novel XR experiences. In order to tap into such prior familiarity, designers need to understand the level of familiarity of potential end users with the technologies they want to employ and adapt design plans accordingly.

10.7 Equity of Access

While we see great potential and opportunities for the use of immersive technologies for disaster preparedness, specifically AR/VR applications, we must also consider critical issues that warrant attention and further research efforts. Several key concerns need to be addressed to ensure comprehensive and socially responsible approaches to the design of immersive environments for disaster preparedness.

First, it is paramount to keep the digital divide and digital inclusion issues front and centre (Dezuanni et al., 2018). Access to high-fidelity technology is not uniform, and there is a risk of exacerbating existing inequalities if community preparedness relies on technology that is only available to some stakeholders. Designs need to cater to different technical literacy levels and varying access to devices and account for the disparities in Internet infrastructure and available bandwidth—as this can hinder the widespread implementation and uptake of immersive technologies. This calls for investigating low-tech and low-fidelity technology alternatives (Gonsalves et al., 2021).

Second, designers need to consider not just urban and metropolitan contexts but also how technical solutions can be transferred to rural and remote communities with precarious Internet access. Research has started to explore the digital capabilities needed for building disaster resilience (Marshall et al., 2023). Failure to address this may prevent vulnerable populations from capitalising on the benefits of advanced disaster preparedness tools.

Third, disaster preparedness often heavily focuses on positivist data sources, neglecting the nuanced ways in which communities make sense of disasters.

Similarly, the community learnings derived from participation in scenarios, AR/VR environments and SGs are of a qualitative nature and can clash with institutional expectations and statutory requirements of local disaster preparedness planning. Research that addresses these epistemological divides is needed to be able to better incorporate community learnings, Traditional Owner knowledge,[2] oral histories and storytelling into disaster planning—recognising the value of both approaches.

Fourth, design scholarship has long recognised the need to emphasise community-centric approaches rather than succumbing to technological solutionism (Morozov, 2014). While we already argued that employing participatory design and co-design is an effective path to capturing the needs and perspectives of end user communities, there remains a risk of governments co-opting technology for solely tokenistic purposes, as mere symbolic gestures or "engagement theatre" (Kamols et al., 2021; Foth et al., 2018). Design "institutioning" refers to an emerging recognition in design research that advocates for meaningful participation and engagement with institutional stakeholders in co-design processes, which can help designers work strategically with the community to guard against superficial uses of technology (Teli et al., 2022).

Designers able to critically examine these issues and successfully navigate equity of access, ability and affordability are more likely to develop designs that account for the social, economic and cultural dimensions of implementing immersive technologies for disaster preparedness.

10.8 Conclusion

This chapter has presented an overview of the ways interactive immersive 3D environments can transform current disaster preparedness practices. With a focus on dynamically evolving extreme events, such as wildfires and severe floods, this chapter argues for the importance of focus on both preparedness and resilience. However, preparedness has often received less attention than resilience, even though investing in preparedness may provide better outcomes for society (Das, 2018). One reason for this lack of attention is the complexity of preparing diverse stakeholder groups, from first responders to entire communities, for evolving, often dangerous situations (Kohn et al., 2013). Nevertheless, the examples canvassed in this chapter demonstrate that an effective broad-based disaster preparedness system can be created. A system with the capacity to immerse people in vicarious, safe, interactive spatial scenarios must be able to engage people in multiple ways, to heighten their experience and reinforce particular behaviours or strategies. SGs use rewards or ratings for this purpose. However, the most effective reinforcement learning approach is founded on personal fulfilment. This arises when a person feels an emotional

[2] This applies to the Australian context as well as to other settler nations in which First Nations' significant knowledge has been marginalised.

connection to the people, places or values they are interacting with in the immersive scenario. Becker et al. (2017) emphasise the importance of emotion in effective experience and preparedness. An important caveat is that, as part of this process, we need to consider how we deal with vicarious or actual trauma, retraumatisation or anticipated potential trauma of users when tapping into previous experiences and in workshopping scenarios or games.

In this chapter, we discussed examples of several different, albeit connected approaches to achieving an effective disaster preparedness environment. From VR and AR to SGs and mock-ups, the chapter presented evidence of the efficacy of these approaches to preparedness and their flexibility and adaptability for accommodating diverse scenarios, narratives and means of visualising extreme events. While the examples in this chapter also reference a range of current technology, more recent developments, such as generative AI, smart glasses and sensory substitution wearables, clearly offer potential for new research into immersive, 3D learning environments that are needed for disaster preparedness. Sharma et al. (2019) demonstrate one way these advances can be leveraged to improve preparedness, incorporating AI agents to simulate complex social dynamics in a collaborative immersive emergency response training. The integration of these emerging techniques alongside VR, AR and SGs is starting to be explored by researchers, and their early findings, such as those in the cited *iFire* project, collectively suggest exciting new opportunities for the field.

Acknowledgements Alethea Blackler's research was supported by the Australian Research Council (ARC) (DP0877964). Nagida Helsby-Clark's PhD research is funded by the ARC (FL200100004, directed by Laureate Prof. Dennis Del Favero) as well as CSIRO's Data 61. Michael Ostwald's research is supported by the ARC (DP220101598 and IC220100030). Marcus Foth's research is supported by the ARC (DP180100174 and LP190100677).

References

Alexander, D. (2000). Scenario methodology for teaching principles of emergency management. *Disaster Prevention and Management, 9*(2), 89–97.

Becker, J. S., Paton, D., Johnston, D. M., Ronan, K. R., & McClure, J. (2017). The role of prior experience in informing and motivating earthquake preparedness. *International Journal. Disaster Risk Reduction* 22 Jun: 179–193, *22*, 179.

Blackler, A., Popovic, V., & Mahar, D. (2010). Investigating users' intuitive interaction with complex artefacts. *Applied Ergonomics, 41*(1), 72–92.

Blomberg, J., & Karasti, H. (2013). Ethnography: Positioning ethnography within participatory design. In J. Simonsen & T. Robertson (Eds.), *Routledge international handbook of participatory design* (pp. 86–116). Routledge.

Das, R. (2018). Disaster preparedness for better response. *International Journal of Disaster Risk Reduction, 31*, 153–159.

Del Favero, D., Thurow, S., Stevens, G., Sharples, J. J., & Davidson, J. W. (2022). Visualizing the unpredictable behavior of wildfires using an artificially intelligent aesthetic. In P. Alsina & I. Vila (Eds.), *Proceedings ISEA 2022* (pp. 696–698). ISEA.

Del Favero, D., Thurow, S., Blackler, A., & Asadipour, A. (2023). Terrestrial aesthetics. Transforming the understanding of and engagement with extreme events. In *Arts and*

Humanities in Digital Transition Conference, Institute of Communication, Nova University of Lisbon, Portugal, 6–7 July 2023.

Dezuanni, M., Foth, M., Mallan, K., & Hughes, H. (Eds.). (2018). *Digital participation through social living labs*. Chandos.

Ehn, P., & Kyng, M. (1991). Cardboard computers: Mocking-it-up or hands-on the future. In J. Greenbaum & M. Kyng (Eds.), *Design at work* (pp. 169–195). CRC.

Ens, B., Goodwin, S., Prouzeau, A., Anderson, F., et al. (2021). Uplift: A tangible and immersive tabletop system for casual collaborative visual analytics. *IEEE TVCG, 27*(2), 1193–1203.

Fischer, S., Itoh, M., & Inagaki, T. (2015). Prior schemata transfer as an account for assessing the intuitive use of new technology. *Applied Ergonomics, 46. PA*, 8–20.

Foth, M., Caldwell, G., Fredericks, J., & Volz, K. (2018). Augmenting cities beyond bedazzlement. In *Workshop Proc. Augmenting cities & architecture with immersive technologies* (pp. 1–4). ACM.

Franklin, B. (1735). *Protection of towns from fire*. Pennsylvania Gazette (February 4): 1.

Gonsalves, K., Foth, M., & Caldwell, G. (2021). Utilizing low-tech AR/VR to engage in communal placemaking during a pandemic. *Interaction Design & Architecture(s), 48*(48), 143–164.

Greenbaum, J., & Kyng, M. (Eds.). (1991). *Design at work*. CRC.

Hatakeyama, H., Nagai, M., & Murota, M. (2016). A flexible scenario-based mobile learning system for disaster evacuation. In *HCI International 2016*. Posters, Toronto.

Hsu, E., Li, Y., Bayram, J., Levinson, D., et al. (2013). State of virtual reality based disaster preparedness and response training. *PLoS Currents, 24*, 5.

IJsselsteijn, W., & Riva, G. (2003). Being there: The experience of presence in mediated environments. In G. Riva, F. Davide, & W. IJsselsteijn (Eds.), *Concepts, effects and measurements of user presence in synthetic environments* (pp. 3–16). IOS.

Kamols, N., Foth, M., & Guaralda, M. (2021). Beyond engagement theatre: Challenging institutional constraints of participatory planning practice. *Australian Planner, 57*(1), 23–35.

Kerr, J., Cheers, J., Gallegos, D., Blackler, A., & Kelly, N. (2023). *The art of co-design: A guide to creative collaboration*. BIS.

Klein, G. (1998). *Sources of power: How people make decisions*. MIT.

Kohn, S., J. Lipkowitz Eaton, S. Feroz, A.A. Bainbridge, …, and D.J. Barnett. 2013. Personal disaster preparedness: An integrative review of the literature. Disaster Medicine and Public Health Preparedness 6.3: 217–231.

Korucu-Kış, S. (2021). Preparing student teachers for real classrooms through virtual vicarious experiences of critical incidents during remote practicum. *Education & Information Technologies, 26*, 6949–6971.

Lazo, J., Bostrom, A., Morss, E., Demuth, J., & Lazrus, H. (2015). Factors affecting hurricane evacuation intentions. *Risk Analysis, 35.10 (Oct)*, 1837–1857.

Lertzman, R. (2019). New methods for investigating new dangers. In P. Hoggett (Ed.), *Climate psychology: On indifference to disaster* (pp. 25–39). Springer.

Lovreglio, R., Gonzalez, V., Feng, Z., Amor, R., et al. (2018). Prototyping virtual reality serious games for building earthquake preparedness: The Auckland City Hospital case study. *Advanced Engineering Informatics, 38*, 670–682.

Ma, D., Shi, Y., Zhang, G., & Zhang, J. (2021). Does theme game-based teaching promote better learning about disaster nursing than scenario simulation: A randomized controlled trial. *Nurse Education Today, 103*. August: 104923, 104923.

Marshall, A., Wilson, C., & Dale, A. (2023). Telecommunications and natural disasters in rural Australia: The role of digital capability in building disaster resilience. *Journal of Rural Studies, 100*, 102996.

Maurer, F., Anslow, C., Jorge, J., & Sousa, M. (2022). Enhancing cross-reality applications and user experiences. In P. Bottoni & E. Panizzi (Eds.), *Proceedings of 2022 International Conference Advanced visual Interfaces* (Vol. 86, pp. 1–3). ACM.

Milgram, P., & Kishino, F. (1994). A taxonomy of mixed reality visual displays. *IEICE Transactions on Information and Systems, E77–D.12*, 1321–1329.

Miller, H. E. (2018). Intuition and decision making for crisis situations. In K. J. Engemann (Ed.), *The Routledge companion to risk, crisis and security in business* (pp. 47–61). Routledge.

Morozov, E. (2014). *To save everything, click here*. Ingram.
Nakanishi, H., Ishida, T., & Koizumi, S. (2009). Virtual cities for simulating smart urban public spaces. In M. Foth (Ed.), *Handbook of research on urban informatics* (pp. 257–269). IGI Global.
Robertson, T., & Simonsen, J. (2013). Participatory design: An introduction. In J. Simonsen & T. Robertson (Eds.), *Routledge international handbook of participatory design* (pp. 1–17). Routledge.
Ryan, M.-L. (2015). *Narrative as virtual reality*. Johns Hopkins UP.
Sánchez, S. Á., Martín, L. D., Gimeno-González, M. Á., Martín-Garcia, T., Almaraz-Menéndez, F., & Ruiz, C. (2016). Augmented reality sandbox: A platform for educative experiences. In F. J. García-Peñalvo (Ed.), *Proceedings of the 4th international conference on technological ecosystems for enhancing Multiculturality* (pp. 599–602). ACM.
Sanders, E. B.-N., & Stappers, P. J. (2008). Co-creation and the new landscapes of design. *CoDesign, 4*(1), 5–18.
Satchell, C., & Foth, M. (2011). Darkness and disaster in the city. *IEEE Internet Computing, 15*(6), 90–93.
Seegert, A. (2009). Doing there vs. being there: Performing presence in interactive fiction. *Journal of Gaming & Virtual Worlds, 1*(1), 23–37.
Sharma, S., Devreaux, P., Sree, S., Scribner, D., Grynovicki, J., Grazaitis, P., Devreaux, P., & Grazaitis, P. (2019). Artificial intelligence agents for crowd simulation in an immersive environment for emergency response. *Electronic Imaging, 31*, 1–8.
Shaw, F., Burgess, J., Crawford, K., & Bruns, A. (2013). Sharing news, making sense, saying thanks: Patterns of talk on twitter during the Queensland floods. *Australian Journal of Communication, 40*(1), 23–39.
Stegmann, K., Pilz, F., Siebeck, M., & Fischer, F. (2012). Vicarious learning during simulations: is it more effective than hands-on training? *Medical Education, 46*(10), 1001–1008.
Still, M. L., & Still, J. D. (2019). Cognitively describing intuitive interactions. In A. Blackler (Ed.), *Intuitive interaction research and application* (pp. 41–61). CRC Press.
Suh, A., & Prophet, J. (2018). The state of immersive technology research: A literature analysis. *Computers in Human Behavior, 86*, 77–90.
Teli, M., McQueenie, J., Cibin, R., & Foth, M. (2022). Intermediation in design as a practice of institutioning and commoning. *Design Studies, 82*, 101132.
Tsai, M.-H., Chang, Y.-L., Shiau, J.-S., & Wang, S.-M. (2020). Exploring the effects of a serious game-based learning package for disaster prevention education: The case of Battle of flooding protection. *International Journal of Disaster Risk Reduction, 43*, 1–9.
Zhao, S. (2003). Toward a taxonomy of copresence. *Presence: Teleoperators and Virtual Environments, 12*(5), 445–455.

Chapter 11
Building Simulations with Generative Artificial Intelligence

Jon McCormack and Mick Grierson

Abstract In this chapter, we explore the possibilities of generative artificial intelligence (AI) technologies for building realistic simulations of real-world scenarios, such as preparedness for extreme climate events. Our focus is on immersive simulation and narrative rather than scientific simulation for modelling and prediction. Such simulations allow us to experience the impact and effect of dangerous scenarios in relative safety, allowing for planning and preparedness in critical situations before they occur. We examine the current state of the art in generative AI models and look at what future advancements will be necessary to develop realistic simulations.

Keywords Augmented reality · Diffusion models · Generative artificial intelligence · Immersion · Machine learning · Simulation · Virtual reality · Visualisation

11.1 Introduction: A Scenario

It is mid-summer in a small rural town in eastern Australia. You are standing on the spacious wooden veranda of your beautiful home, surveying an expansive vista of wild flora nestled in the valley formed by two distant mountains. It is a clear, sunny day. You feel a hot, dry, gusty wind on your face. All around you, the landscape is parched—heavy rains earlier in the year allowed the surrounding landscape to

J. McCormack (✉)
SensiLab, Monash University, Caulfield East, VIC, Australia
e-mail: jon.mccormack@monash.edu

M. Grierson
Institute for Creative Computing, University of the Arts London, London, UK
e-mail: m.grierson@arts.ac.uk

D. Del Favero et al. (eds.), *Climate Disaster Preparedness*, Arts, Research, Innovation and Society, https://doi.org/10.1007/978-3-031-56114-6_11

blossom, and it filled with tall grasses and a rich variety of tinder. But a few months ago, the rains gave way to continuous weeks of well-above-average temperatures. Rain has not fallen in months, and now everything is a faded hue of off-yellow and completely dry.

As you turn to go inside, you hear a far-off roar that sounds like the rumble of thunder. The sound seems to be getting louder. As the wind picks up, you notice an orange glow on the horizon. Within the space of just a few minutes, it becomes quite clear that a massive bushfire is bearing down on you. Thousands of tiny burning embers are blowing around you, mixed with an acrid, grey smoke that chokes the atmosphere, soon making it difficult to see more than a few metres in any direction. The pleasant blue sky has quickly turned pitch black, creating an eerie sense of disquiet.

As the wall of super-heated flames—a minute ago just a pretty deep orange glow in the distance—are now closing in on you and your property at over 100 km/hour, you have just a few seconds left before the wall of fire will consume you. You see several animals running just ahead of the fire front, but less than a second later, they are swallowed by the fire and disappear. A sense of panic and dread kicks in as it feels as though the blood is draining from your body. You rush out to the back of the property to locate the large steel door that leads down to an emergency fire bunker. You pull the vegetation that has grown around the door away as the roar becomes unimaginably loud; the thick black smoke has turned day into the darkest night. You scramble to get into the bunker, pulling the heavy steel door shut, just as an enormous wave of fire engulfs your home and land. You sit inside the dark bunker, the intense roar of the fire still audible, and a wave of super-heat can be felt above. You notice your heart is racing and pounding against the wall of your chest. You take deep breaths, telling yourself to stay calm while trying to convince yourself that you will survive.

A few moments later, the attendant removes the virtual reality headset and bodysuit you were wearing, letting you know that "the simulation is over", reassuring you that "you are safe now and there's nothing to worry about". You suspect that you must look very frightened and distressed as you are asked if you need a few minutes to take in what you have just experienced, before completing your disaster readiness training and heading back home. You are in your town's local community hall, and it is only a short drive back to your property—the same one that you just experienced burning to the ground with a realism so visceral you are now sweating profusely and in a mild state of shock. This is not an experience you will quickly forget.

11.1.1 Building Simulations

While the scenario just presented is currently largely in the realm of speculative fiction, the technology to produce such a simulation has made rapid advances over the last few years, suggesting that it may shift from speculation to practical realisation in the next decade. Further, in addition to being able to provide virtual reality (VR) simulations of familiar environments for training purposes, it is potentially possible that such experiences could even be simulated with generative AI using augmented reality (AR) in situ. Currently, however, building realistic simulations for immersive technologies—such as VR and AR—is a complex and time-consuming process.

Three-dimensional (3D) simulations usually begin with modelling the geometry and textures of every object that will appear in the simulation. Despite advances in 3D modelling techniques and the wide availability of existing models online, this is a highly specialised and time-consuming task. Beyond the modelling of physical form, a simulation also needs to model behaviour. For the simulation to be realistic, that behaviour must be accurate to reality or at least be plausible to reality. Lastly, for the simulation to be immersive, it must provoke a strong sense of presence, convincing any participants that what is happening is "real". This typically involves high-fidelity images, sound, haptics, kinaesthetics and beyond.

Our speculative scenario made use of sensory experience beyond mere visual simulation, incorporating sound, proprioception, kinaesthetic, haptic and even olfactory simulation (e.g. the smell of smoke). Not all of these sensory modalities are currently well synthesised by AI, and a heavy commercial focus on visual and audio synthesis currently dominates the well-known foundational AI models (Bommasani et al., 2021).

11.2 Simulation of Extreme Event Scenarios

In this section, we look at the current state of the art in generative AI and how it might be usefully purposed for immersive simulation.

As we noted elsewhere, "over the last decade, a [number] of innovations in generative machine learning (ML) models have allowed the generation of photo-realistic images of [nonexistent] people (Karras et al., 2018), coherent paragraphs of text (Vaswani et al., 2017), conversion of text directly to [runnable] computer code and, [more] recently, from text descriptions to images (Ramesh et al., 2022), video (Singer et al., 2022; Blattmann et al., 2024), and 3D models (Gao et al., 2022)" (McCormack et al., 2023). Neural radiance fields (NeRFs) (Mildenhall et al., 2020) can synthesise 3D scenes from novel viewpoints using sparse 2D images as input and guided by text descriptions (Zhang et al., 2023). Tools such as these are already being offered to creators through platforms such as NVIDA's Open USD-based Omniverse.

These tools are increasingly used in audio-visual production, combining a range of generative AI techniques, including diffusion models (Yang et al., 2023), specialised generative adversarial networks (Iglesias et al., 2023), autoencoders and image-to-image systems (Wang et al., 2018). Initially popular for their potential for still image generation, they have more recently become surprisingly usable for video and 3D scene generation. As noted elsewhere, "systems such as DALL-E 2, MidJourney and Stable Diffusion allow the generation of detailed and complex imagery from short text descriptions. These text-to-image (TTI) systems allow anyone to write a brief description [(a 'prompt')] and have the system respond with a series of images that depict the scene described in the text, typically within 5 [to 30] seconds" (McCormack et al., 2023). More recently, diffusion model-based text-to-image systems have demonstrated rapid advances in both quality and popularity. At the time of writing, these systems can produce high-quality imagery as fast as a person can type in a prompt (Stability.ai, 2023). They can also facilitate image editing and manipulation.

As my team noted, "the obvious source of these systems' popularity is that they offer something new: being able to generate an image, [video sequence or 3D render] just by describing it, without having to go to the trouble of learning a skill—such as [illustration,] painting, photography, [cinematography or 3D modelling]—to actually make it. And importantly, the quality and complexity of the [media] generated is often [comparable] to what an experienced [professional] human creator could produce, [at least at the surface level. Moreover, generative AI] systems demonstrate a semantic [interpretation] of the input text and can convert those semantics so that (in some cases) they are more-or-less coherently represented in the generated images. This new-found capability has inspired many useful image generation and [manipulation possibilities,] such as 'outpainting', where a pre-existing image can have its edges [extended] with coherent and plausible content, or as an 'ideation generator', where new versions of a set of input images are generated" (McCormack et al., 2023).

11.2.1 Use in Visual Simulation

The idea that through new generative AI technologies we can construct high-fidelity simulations of real-world events presents a step change in developing simulation systems. Rather than labouring over detailed 3D models, building complex simulations by hand or using digital media such as cinema or photomedia to construct a rich simulation experience, generative AI potentially presents the opportunity to deliver high-fidelity simulations simply by describing them in language.

Current text-to-image (TTI) systems rely on diffusion models. These models are trained by adding noise to a training set, forcing the model to learn how to convincingly reconstruct image representations. This approach has significant advantages in image generation quality over previous methods such as generative adversarial networks (GANs) (Goodfellow et al., 2020). The fundamental innovation of TTI systems lies in the integration of two different approaches—a language transformer model that accepts image descriptions as text and an image generator that synthesises the image. The transformer is usually based on CLIP (Contrastive Language–Image Building Simulations with Generative AI 5 Pre-training) models, a neural network that learns visual concepts from natural language supervision (Radford et al., 2021). This is a significant improvement over previous models such as convolution neural networks (CNNs), which excelled at basic classification of objects in

Figs. 11.1 and 11.2 Images of bushfire created using *Stable Diffusion*

an image but could not recognise more salient concepts such as style, context or semantics. The image generator uses a multi-step process that operates in the image latent space, using a UNet neural network and scheduler. The output of this "diffusion" process is an image tensor that is decoded into an image by an autoencoder.

To illustrate the potential of these systems for simulation, we used an open-source version of Stable Diffusion. Figures 11.1 and 11.2 show two sample AI-generated images created using Stable Diffusion. The prompts used were "national geographic photo of an Australian bushfire, landscape, trees" (left) and "national geographic photo of fire-fighters with a hose fighting a large bushfire" (right). As can be seen, the prompts generate quite "realistic" images that would typically be associated with Australian bushfires. Using phrases such as "national geographic photo" pushes the system into producing high-quality, documentary-like images, as would typically be associated with *National Geographic* (we could have specified "old black and white daguerreotype" or "Banana Fish anime" to completely change the aesthetic style of the image).

This simple example highlights some of the issues with creating prompts: that one needs to be quite specific in the prompt about details such as surface aesthetics, style, context, etc. Such a requirement leads to much of the prompt language containing references to the visual aesthetics of the image: including style, lighting, level of detail, even descriptions of camera lens focal lengths, angle or position of the shot and other various cinematic conventions. The necessity of providing such detail on surface aesthetics, composition, etc. mirrors CLIP's ability to capture these image qualities as general image features irrespective of the objects depicted in them.

Figure 11.3 shows another example of generative AI simulation of flooding events. To create these images, the following prompts were used: "national geographic photo of people piling sandbags in an Australian town after flooding" (left) and "national geographic photo of an Australian town after mild flooding" (right). In this example, the prompts are interpreted correctly, but only to a point. The way the sandbags are being piled does not really make practical sense (they would be unlikely to mitigate the effects of rising water), the "people" depicted do not have

Figs. 11.3 and 11.4 Images of flooding created using Stable Diffusion

natural poses or body parts, the layout of the town is impossible, and so on. An obvious solution to these issues would be to provide more specific directions.

However, providing too much detail on the objects and their relations in the scene does not necessarily translate into the generated image. Using Fig. 11.1 as an example, if we were to modify the prompt used in the second image to be more specific about the exact number of firefighters, specifics of their individual poses, details of their uniforms or their specific location in the image, the results are unlikely to directly match the prompt. For example, if we specified "five firefighters", we may get any number between two and ten or more. This is due to the way CLIP converts the input prompt into a latent embedding.

Another potential approach is to use an image-to-image (ITI) method. These are largely similar to TTI systems, and most popular diffusion-based TTI platforms offer an ITI mode. Using this approach, imagery can be adapted using a text prompt to make modifications and transformations while preserving important characteristics of the scene. There are a few potential methods for generating realistic scenes more easily using ITI. For example, simple 3D environments can be created using existing 3D models and then modified using generative AI to render more realistic environments for simulation. This is demonstrated in Fig. 11.3 using Stable Diffusion. A basic 3D environment is transformed into a more complex scene through the use of a text prompt while preserving the overall structure and characteristics of the scene. There are a few potential problems with this approach. For instance, it can be a challenge to control the content that might appear in the generated scene due to the diversity of images used to train the model. In Fig. 11.3, the generated image features a boat in place of a shadow, which appears in the input image. This can be mitigated by adjusting the strength of the transformation. Another approach is to use fine-tuning methods, including the creation of custom low-rank adaptors (Hu et al., 2021) to guide content generation more explicitly with examples and custom embeddings. Despite offering considerably more control, problems with content consistency still cannot be entirely avoided with approaches such as these (Figs. 11.5 and 11.6).

Aside from adjusting and controlling content with prompting, there are potentially other, more direct content control methods that may be more practical. One

Figs. 11.5 and 11.6 A low-quality 3D render is shown on the left, with an image-to-image version on the right generated by Stable Diffusion. The ITI prompt was simply "A flooded art gallery"

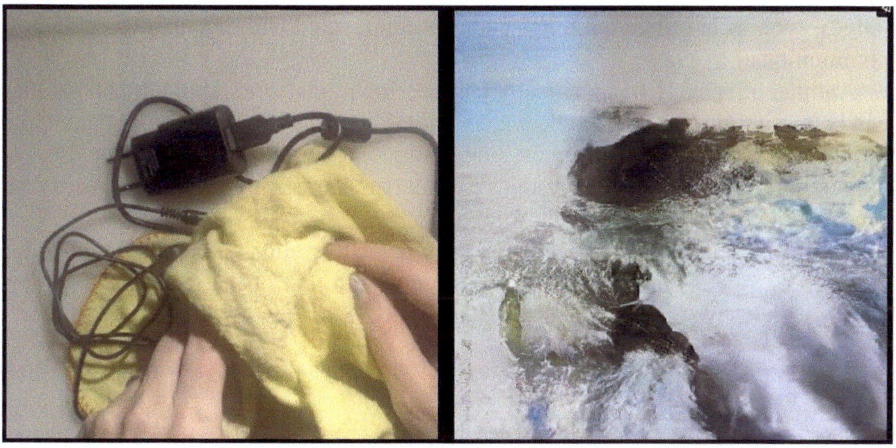

Fig. 11.7 Image from *Learning to see: Gloomy Sunday, 2017*. A simple ITI approach is used to create a real-time transformation of webcam input. The input image on the left-hand side is transformed into imagery of the sea and rendered in under 20 milliseconds, allowing for interactive control of the generative AI system

could use specifically curated models, such as fire and/or water generators. These models could be specifically trained to add photo-real fire and water effects on input images using ITI methods, similar to the approach taken in Memo Akten's Learning to See: Gloomy Sunday (Akten et al., 2019), as shown in Fig. 11.7. Here, models were created from datasets of open water, fire, flowers and other categories of images, and these were then used for generation.

Using this approach, models need not understand a wide range of different kinds of imagery, have knowledge of context, nor draw on complex text prompts in order to guide generative image models. Instead, input images could be adapted by models with highly limited yet detailed and carefully engineered outputs. These kinds of models are far simpler than large, monolithic text-guided diffusion systems and as a result can be run in real time in high definition on modern hardware. Furthermore, as these models could be specifically tailored to the problem of disaster simulation, they could also contain information on behaviour, for example, through the application of fluid dynamics models within the content generation pipeline.

This approach could be applied to the generation of content for simulations in AR environments. Focused, bespoke generative AI models such as those based on ITI approaches described above could quite easily be deployed in order to transform real-time stereoscopic image feeds from AR headset cameras, with the transformed output rendered directly to the headset in real time. This creates the opportunity for generating disaster simulations in real-world environments, where participants experience photo-real 3D generative AI simulations superimposed on the world as it exists. This requires less computing power as only the transformed elements need rendering. In addition, it could provide the opportunity to conduct disaster readiness training in situ with greater fidelity, allowing firefighters and other emergency services personnel to model scenarios in specific locations where there are known challenges, such as in public buildings and city centres and with potentially vulnerable communities.

Another advantage of specifically engineering models for simulation is that they can be more transparently usable and explainable than currently popular contemporary text-guided diffusion systems. For example, it can be challenging for users of contemporary generative TTI systems to understand precisely which aspects of their prompt may be having the greatest impact.

As illustrated in Table 11.1, human prompt writers often tend to over-equate the complexity, poetics and quality of the prompt with that of the resultant generated

Table 11.1 An example of differences between human-generated prompts and machine-based description

Prompt	Generated image	Description
Imagine a dream-like scene where reality blurs and the boundaries between woman and peacock dissolve. Sketch a woman's body full of delicate vulnerability, her features soft and poetic. Let the peacock's head emerge, seamlessly integrating with its essence, symboli-sing the deep connection with the world of colours of the peacock's tail. Use the impasto technique to add a tactile quality, allowing the viewer to visually feel the texture of the artwork. Set against a deep, velvety canvas of dark blue on a black background, this ethereal combination creates a sense of enchantment, encouraging viewers to explore the depths of their imagination		Painting of a woman with peacock feathers on her head

image. The table shows a relatively long human-authored prompt (left) and the resultant image generated by Midjourney (middle). We ran this image through a state-of-the-art BLIP (Bootstrapping Language-Image Pre-training for Unified Vision-Language Understanding and Generation) image captioning model (Li et al., 2022), which can generate a descriptive caption for any image. The captioning model gives an overall description of the image, much like the way a human would, as the model recognises not only the objects in a scene but also the relationships and basic surface aesthetic properties. Table 11.1 (right) shows the results of this "machine eye" perception of the image. As can be observed, this is a far more direct and literal description of the image than that of the original prompt.

11.3 Accuracy and Ethics

In addition to issues of transparency and explainability in contemporary TTI systems, there are considerable problems with both the accuracy and also the ethical grounding of many large foundation model-based generative AI tools. Such systems are not being developed within the constraints of well-defined use cases, nor with domain building simulations with generative AI-specific requirements in mind. Evidence points to the need for the development of generative AI that is specifically tailored to the problems of simulation.

For example, careful examination of the images shown in Figs. 11.1, 11.2, 11.3 and 11.4 reveals problems of accuracy and bias. Due to both the data used for training and the nature of diffusion systems, references such as "Australian landscape" tend to get translated into cliched representations based on statistical averages in the training data. A number of studies have analysed AI-generated images, identifying a wide range of biases, such as under-representing certain race groups (Bansal et al. 2022; Naik & Nushi, 2023), cultural gaps (e.g. over-representing specific nations (Naik & Nushi, 2023), or the reinforcement of stereotypes (e.g. "a photo of a lawyer" consistently showing a white male) (Bianchi et al., 2023).

A recent analysis of 3000 images generated by Midjourney using prompts to depict national identities also highlighted tendencies towards bias and stereotypes prevalent in generative AI systems. For example, prompting an image of "New Delhi's streets" generated images that were mostly portrayed as polluted and littered (Turk, 2023). This perpetuates cultural norms that are prevalent in training datasets while under-representing less stereotypical and non-Western aspects of culture, society and landscape. Although some researchers have proposed ways to mitigate these effects, such as adding specific phrases, e.g. "irrespective of gender" (Bansal et al., 2022), or through the use of more specific prompts to mitigate bias, these mitigation strategies are often ineffective (e.g. despite explicitly mentioning words such as "white", "wealthy" or "mansion", Bianchi et al. (2023) report that Stable Diffusion continues to associate poverty with people of colour).

There is emerging research exploring ways that AI systems can be potentially better designed through the inclusion of those with domain expertise in fields where models will be deployed. Co-production of ML systems is a developing international field that attempts to respond to risks including those mentioned above. Recent work (e.g. Grabe et al., 2022) indicates existing research on AI system design does not adequately address design challenges posed by AI. They propose a method for understanding the potential complexities of design through two specific features: uncertainty regarding system capability, as exemplified by the lack of system transparency highlighted above, and output complexity, which, as we have described, is a fundamental problem for TTI approaches using foundation models. Other work (e.g. Mucha et al., 2020) highlights the importance of creating AI interfaces tailored to users' needs and of gaining feedback from users early in the design process, supporting the fundamental principle that generative AI systems for simulation should be specifically designed through collaboration with domain experts and that this approach is vastly preferable to the use of existing general-purpose TTI systems in the context of simulation design.

11.3.1 Data Laundering

As we noted elsewhere, "one of the key factors that contributes to the capability of TTI models is their access to [very large] datasets used for training and validation. Achieving the visual quality and diversity that they are capable of reproducing requires a [vast] corpus of human-created imagery, which is typically scraped from the internet, in a practice that has been dubbed 'data laundering'. Scraped datasets—which [may] include copyrighted media—rely on special exemptions for 'academic use' to avoid any legal barriers preventing their use, or for copyright owners to claim against (Baio, 2022). For [example], Stability AI (the creators of Stable Diffusion) funded the Machine Vision & Learning research group at the Ludwig Maximilian University of Munich to [undertake] the model training and a small [not-for-profit] organisation, LAION, to create the training dataset of approximately 5.85 billion images, many of which are [copyrighted], and in general appropriated for this purpose without the image [creators'] direct permission" (McCormack et al., 2023).

We further noted that artists have "raised [concerns] about the ethical and moral implications of their work being used in such systems. These concerns include the appropriation of an [individual] artist's 'style', mimicry, and even the replacement of a [specialist] human artist or illustrator. Furthermore, there is [currently] no easy way to be excluded or removed from such datasets" (McCormack et al., 2023), and any mechanisms are generally "opt-out", meaning that unless you take action to prevent your own data from being excluded, it is considered fair game for scraping. The use of copyrighted material in AI training data is currently being tested legally in several different countries. Governments may need to draft new legislation to deal with these issues, as has already happened in the European Union.

11.3.2 Copyright Issues

The use of copyrighted images in datasets highlights the question of whether training models on copyrighted data should be considered plagiarism or a form of copyright infringement. As we pointed out elsewhere, "being able to easily generate an image in [a specific] artist's or [house] style [(e.g. 'National Geographic')] without paying for that artist to create it (or paying any royalties or licensing fees), allows users of such technology to bypass the traditional economic, legal and moral frameworks that have supported artists and businesses traditionally. Generating copyright-free images immediately for commercial use without the cost or time involved in securing copyright from a [human] artist may become an attractive proposition, raising the interesting legal [question of who would be the defendant in any copyright infringement case brought about by this scenario" (McCormack et al., 2023).

11.3.3 Making AI "Safe"

Beyond ethical questions involving the sources of data and representational bias are the mechanisms by which many large companies try to ensure that generative AIs are "safe". Many models are augmented with what is known as "Reinforcement Learning from Human Feedback" (RLHF), where outsourced workers in developing economies are paid to "sit for several hours every day watching videos of harmful content and analyzing textual descriptions of hate speech, sexual violence, bestiality, and violence" (Ngila, 2023). This human tagging or classifying of unsafe content is used to train additional AIs that filter results to prevent the underlying generative system from showing harmful content.

Some people have already developed psychological dependencies or been prompted to take real-world action following advice from generative AI systems, with both positive and negative results, including suicide, divorce or self-harm. As models become even more sophisticated, we are likely to see new forms of human–AI relationships with potentially dangerous results. In the context of simulation, there are a number of important considerations for the simulation to be credible. Generative AI suffers from what is euphemistically referred to as "hallucinations"—factually incorrect or erroneous results. The implications for a generative AI "hallucinating" in a simulation context can range from benign to catastrophic, depending on context and situation. For example, the "Australian town" depicted in Fig. 11.4 does not exist, and no real town would be structured in the way it is depicted. Simulations may be speculative, allowing us to ask, "what if…", but if the answer is based on factual inaccuracies, the value of the simulation may be worthless.

11.4 Conclusion

These issues, when considered in the round, support the overall conclusion that specific, tailored bespoke generative AI models could offer significant advantages over large, monolithic generative AI tools in the context of disaster simulation. They are more controllable in terms of content, as the training process can effectively constrain their output to a known selection of imagery labels. They are more efficient, being able to transform multiple high-resolution video streams in real time on a single modern computer having a relatively modest capacity. They could be used to augment existing, low-quality 3D scenes with photo-realistic real-time generative AI incorporating relevant and plausible behaviour. They could also be used to render stereoscopic photo-real AR experiences for emergency readiness training in real-world environments. They are potentially more transparent, being trained on known data that could be specifically selected by domain experts. As a result, they are less likely to generate out-of-domain hallucinations, potentially offering greater accuracy, avoiding potential copyright infringement and mental health risk to those working with them.

11.4.1 Limitations: Multimodal AI

In this chapter, our main focus has been on exploring the use of AI to generate realistic imagery for visual simulation. As discussed, many contemporary systems use text prompts to generate output—e.g. text to image, text to video and text to 3D model—and in these ways are not too distinct from a text-based search. We have also explored how more bespoke generative AI systems can potentially play a significant role in the future. However, an obvious current limitation of this analysis is that the interaction with such generative AI systems is uni-modal.

However, as our simple scenario in Sect. 1 demonstrates, an immersive simulation is a multimodal experience, encompassing multiple senses and ways of interacting. Multimodal interaction has been well studied from a human–computer interaction perspective (see, e.g. McCormack et al., 2018 for an overview). Recently, multimodal generative AI systems have been gaining traction. These systems consider multiple modes of input and output (e.g. text, image, video) allowing cross-modalities to be considered. For example, Google DeepMind recently announced a new AI platform, which they call "Gemini" that allows "reasoning seamlessly across text, images, video, audio, and code" (Google DeepMind, 2023). While still in development, multimodal AI systems have the potential to analyse scenes or environments and to then ask questions that would require expertise ("how safe would this exit be in a fire?", "where is the safest place to go if this area is under imminent threat of flooding?"). It may be that these multimodal systems are better able to generalise scenarios as a result of constructing representations from a greater number of dimensions, for example, combinations of sound, image and text. The capability and power of such systems is potentially enormous, and more research is needed in order to understand how they might one day be deployed for the purposes of simulation.

References

Akten, M., Fiebrink, R., & Grierson, M. (2019). *You are what you see*. SIGGRAPH.

Baio, A. (2022). AI data laundering: How academic and nonprofit researchers shield tech companies from accountability. Waxy.org. https://t1p.de/lssgi. Accessed 15 Dec 2023.

Bansal, H., Yin, D., Monajatipoor, M., & Chang, K. W. (2022). How well can text-to-image generative models understand ethical natural language interventions? In Y. Goldberg, Z. Kozareva, & Y. Zhang (Eds.), *Proceedings of Conference on Empirical methods in natural language processing* (pp. 1358–1370). ACL.

Bianchi, F., Kalluri, P., Durmus, E., Ladhak, F., et al. (2023). Easily accessible text-to-image generation amplifies demographic stereotypes at large scale. In *Proceedings of ACM Conference on Fairness, accountability and transparency* (pp. 1493–1504). ACM.

Blattmann, A., Dockhorn, T., Kulal, S., …, & Rombach, R. (2024). *Stable video diffusion: Scaling latent video diffusion models to large datasets*. https://t1p.de/mj6wf. Accessed 15 Dec 2023.

Bommasani, R., Hudson, D., Adeli, E., Altman, R., Arora, S., …, & Liang, P. (2021). *On the opportunities and risks of foundation models*. Arxiv.org. https://t1p.de/gcl9r. Accessed 15 Dec 2023.

Gao, J., Shen, T., Wang, Z., Chen, W., et al. (2022). Get3d: A generative model of high quality 3D textured shapes learned from images. *Advances in Neural Information Processing Systems*. Arxiv.org. https://t1p.de/ji55g. Accessed 15 Dec 2023

Goodfellow, I., Pouget-Abadie, J., Mirza, M., Xu, B., et al. (2020). Generative adversarial networks. *Communications of the ACM, 63*(11), 139–144.

Google. (2023). Welcome to the Gemini era. Google. https://t1p.de/06u8v. Accessed 15 Dec 2023.

Grabe, I., Duque, M., Risi, S., & Zhu, L. (2022). Towards a framework for human-AI interaction patterns in co-creative GAN applications. In *IUI Workshops* (pp. 92–102) SemanticScholar.org. https://t1p.de/21dkt. Accessed 15 Dec 2023

Hu, E. J., Shen, Y., Wallis, P., Allen-Zhu, Z., …, & Chen, W. (2021). *Lora: Low-rank adaptation of large language models*. Arxiv.org. https://t1p.de/berdx. Accessed 15 Dec 2023.

Iglesias, G., Talavera, E., & lvarez, A. (2023). A survey on GANs for computer vision: Recent research, analysis and taxonomy. *Computer Science Review, 48*, 100553.

Karras, T., Laine, S., & Aila, T. (2018). *A style-based generator architecture for generative adversarial networks*. Arxiv.org. https://t1p.de/cjn6s. Accessed 15 Dec 2023.

Li, J., Li, D., Xiong, C., & Hoi, S. (2022). *Bootstrapping language-image pre-training for unified vision-language understanding*. https://t1p.de/jvybx. Accessed 15 Dec 2023.

McCormack, J., Roberts, J., Bach, B., Freitas, C., et al. (2018). Multisensory immersive analytics. In K. Marriott, F. Schreiber, T. Dwyer, K. Klein, et al. (Eds.), *Immersive analytics* (pp. 57–94). Springer.

McCormack, J., Cruz Gambardella, C., Rajcic, N., Krol, S. J., et al. (2023). Is writing prompts really making art? In C. Johnson, N. Rodríguez-Fernández, & S. M. Rebelo (Eds.), *AI in music, sound, art and design* (pp. 196–211). Springer.

Mildenhall, B., Srinivasan, P., Tancik, M., …, & Ng, R. (2020). *Nerf: Representing scenes as neural radiance fields for view synthesis*. Arxiv.org. https://t1p.de/zmsgb. Accessed 15 Dec 2023.

Mucha, H., Robert, S., Breitschwerdt, R., & Fellmann, M. (2020). *Towards participatory design spaces for explainable AI interfaces in expert domains*. Fraunhofer Institut. https://t1p.de/62e5k. Accessed 15 Dec 2023

Naik, R., & Nushi, B. (2023). Social biases through the text-to-image generation lens. In *Proceedings of AAAI/ACM Conference on AI, ethics & society* (pp. 786–808). ACM.

Ngila, F. (2023). OpenAI underpaid 200 Kenyans to perfect ChatGPT—Then sacked them. *Quartz*. https://t1p.de/ftdpc. Accessed 15 Dec 2023
Radford, A., Kim, J., Hallacy, C., …, & Sutskever, I. (2021). *Learning transferable visual models from natural language supervision*. Arxiv.org. https://t1p.de/z25q7. Accessed 15 Dec 2023.
Ramesh, A., Dhariwal, P., Nichol, A., Chu, C., & Chen, M. (2022). *Hierarchical text-conditional image generation with clip latents*. Arxiv.org. https://t1p.de/12kic. Accessed 15 Dec 2023.
Singer, U., Polyak, A., Hayes, T., Yin, X., …, & Gafni, O. (2022). *Make-a-video: Text-to-video generation without text-video data*. Arxiv.org. https://t1p.de/bex23. Accessed 15 Dec 2023.
Stability.ai. (2023). Introducing sdxl turbo: A real-time text-to-image generation model. *Stability AI*. https://t1p.de/k1xqx. Accessed 15 Dec 2023.
Turk, V. (2023). *How AI reduces the world to stereotypes*. Restofworld.org. https://t1p.de/9rm40. Accessed 15 Dec 2023.
Vaswani, A., Shazeer, N., Parmar, N., Uszkoreit, J., et al. (2017). Attention is all you need. *Advances in Neural Information Processing Systems, 30. Nips.cc*. https://t1p.de/70qm4. Accessed 15 Dec 2023
Wang, T., Liu, M., Zhu, J., Liu, G., et al. (2018). Video-to-video synthesis. *Advances in Neural Information Processing Systems*. https://t1p.de/mm6ua. Accessed 15 Dec 2023
Yang, L., Zhang, Z., Song, Y., Hong, S., …, & Yang, M. (2023). *Diffusion models: A comprehensive survey of methods and applications*. Arxiv.org. https://t1p.de/zx5sp. Accessed 15 Dec 2023.
Zhang, J., Li, X., Wan, Z., Wang, C., & Liao, J. (2023). *Text2nerf: Text-driven 3D scene generation with neural radiance fields*. Arxiv.org. https://t1p.de/339xt. Accessed 15 Dec 2023.

Chapter 12
Rehearsing Emergency Scenarios: Using Space Syntax and Intelligent Mobility Modelling for Scenario Visualisation and Disaster Preparedness

Michael J. Ostwald and S. Travis Waller

Abstract Extreme climate events require people to rapidly navigate dynamically changing environments. Wildfires and floods alter the landscape, blocking roads, destroying landmarks and turning the built environment and infrastructure into potential hazards. While various computational methods exist for modelling the ways people move through buildings, urban spaces and transportation networks, there are relatively few examples of these being applied to natural disasters. Moreover, these methods have unexploited potential to support real-time simulation and visualisation of the evolving impacts of climate emergencies. This chapter reviews advanced research using two computational approaches—space syntax and intelligent mobility modelling (IMM)—to visualise the interaction between people, the built environment and infrastructure. These approaches support the simulation of diverse scales of spatial interactions, from individuals to entire populations. Combining examples from the authors of research in these fields with practices and concepts from the arts, this chapter highlights the ways new applications of these methods can support stakeholders's needs for disaster responsiveness, rehearsal and preparedness.

Keywords Data visualisation · Intelligent mobility modelling · Natural disasters · Network dynamics · Space syntax

M. J. Ostwald (✉)
School of Built Environment, The University of New South Wales, Kensington, NSW, Australia
e-mail: m.ostwald@unsw.edu.au

S. T. Waller
Institut für Verkehrsplanung & Strassenverkehr, Technical University of Dresden, Dresden, Germany
e-mail: steven_travis.waller@tu-dresden.de

D. Del Favero et al. (eds.), *Climate Disaster Preparedness*, Arts, Research, Innovation and Society, https://doi.org/10.1007/978-3-031-56114-6_12

12.1 Introduction

This chapter explores the capacity of two research approaches—space syntax and intelligent mobility modelling (IMM)—to improve the ways people prepare for climate emergency challenges. Wildfires and floods change the environment in unpredictable and catastrophic ways. During these events, the places where people would typically find shelter and the paths they routinely take to get there can no longer be relied upon. Buildings can be destroyed by fire, roads and rail lines can be blocked by flood, and the environment can be obscured by smoke and debris. Both wildfires and floods disrupt people's understanding of space, undermining their ability to navigate it safely. This is why there is a need to prepare people for dynamically changing environmental conditions to increase survival in the face of climate emergencies. To borrow a concept from the arts, effective *preparedness* for such a situation requires *practice* and *rehearsal* before the event. But how can advanced spatial research support this goal?

Space syntax and IMM are used for a range of purposes, from simulating types of human behaviour to predicting the impact of movement-related decision making. This chapter considers how these approaches can be used to immerse people in spatial data, allowing them to experience, practice, rehearse and perform appropriate responses to wildfires and floods. This goal reflects the growing realisation that the visualisation strategies used in the arts and in design are effective ways of making advances in socio-spatial modelling more accessible, immediate, engaging and compelling for the general public (Gattenhof et al., 2021; Ludlow & Travis, 2018). Proposing an arts-based approach to translating data and methods from space syntax and IMM into an immersive rehearsal space is possible precisely because they have a common foundation in two factors: space and cognition.

The capacity to develop knowledge about an environment, understand it and then use it strategically is a spatio-cognitive skill. A subset of cognitive psychology, spatial cognition encapsulates the ways people acquire, organise and then apply spatial information. As such, spatial cognition shapes the ways people traverse, explore and locate themselves in the world. People use spatial cognition to commute to work, move about their workplace and decide where to meet colleagues or friends. Collectively, this shapes the volume of traffic on a motorway, the number of people using public transport or the ways crowds leave a concert. As such, while it is individuals who employ spatial cognition, the impacts of large-scale spatio-cognitive patterns shape human society.

The combination of space and cognition into a rigorous model is possible because they have measurable or observable characteristics. This means that researchers can model the relationship between environmental characteristics on the one hand and human behaviours and responses on the other. For example, space has both measurable "geography" (location, length, breadth, height) and "topology" (connectedness, accessibility, control) (Hillier & Hanson, 1984). For many decades, empirical researchers have identified patterns connecting human behavioural responses to spatial properties. Some of these patterns relate to movement or wayfinding, which can be statistically generalised to predict large-scale human activities (Lee & Ostwald, 2020). Others relate to perceptual (directionality, intelligibility, complexity) or emotional (security, exposure, confusion) responses to space, which can also be generalised using a range of methods, from statistical correlation (Ostwald & Dawes, 2018) to machine learning (ML) (Jalalian et al., 2011; Wong et al., 2012).

Multiple fields have developed research connecting the spatial properties of environments and human cognition and behaviour using models or simulations. Two of these fields, which have a common interest in movement and navigation patterns, are the focus of the present chapter. However, while space syntax and IMM can simulate human behaviours or transport choices, their conventional applications are often to networks of spaces or streets with known or predictable properties. This raises the question, what about those situations where this is not the case? For example, can they be used to better understand, visualise and prepare people to make spatial decisions about escaping wildfires or floods or evacuating a city shelled by artillery or targeted by drones?

This chapter commences with an overview of the types of thinking and modelling involved in space syntax and IMM and their application in response to dynamic environmental events, from natural disasters to warfare. The section includes several examples of the transformation of spatial data into immersive simulations. The value of these examples is that they suggest a pathway for translating the findings of advanced spatial methods—typically supported by ML and artificial intelligence (AI)—so that they can communicate more directly to people through environmental simulations, gaming and immersive educational experiences.

The four-part process required for this translation to occur involves developing the capacity to (i) *picture* or visualise extreme events, (ii) *narrate* or dramatise them, (iii) *rehearse* or test responses and finally (iv) *communicate* how we can better prepare for them. This chapter is focused on the third process, rehearsal, where environmental and behavioural data are combined and visualised to prepare people for extreme events. As such, the penultimate section of the chapter explains the idea of a rehearsal and considers how the arts may employ methods such as space syntax and IMM to improve disaster preparedness and resilience in society. This chapter concludes with examples from the authors, which can be developed in future research. Ultimately, this chapter is about how the arts, and in particular modes of practice, rehearsal and performance, can provide a pathway for big spatial data to prepare people for unpredictable climate events.

12.2 Background to Spatial Analytics

Both space syntax and IMM view human behaviours as operating in spatial networks. As an example, at a smaller scale, it is possible to create a network diagram, or graph, of the ways rooms in a house are connected. This network, called a "permeability graph", encapsulates the social patterns of the people who designed and constructed the house. Similarly, the roads that make up the neighbourhood around the house can be converted into a network or graph of the spaces people will move through to go from one location to another. This can be expanded to create a network of all major city roads, rail lines and cycleways. Each of these networks can be modelled mathematically to explore, for example, the extent to which certain street intersections (network nodes) are critical to traffic flow in a city. At a neighbourhood scale, mathematical analysis of the network of spaces can be used to determine the most critical areas for wayfinding, signage or visual landmarks. While these general principles connect space syntax and IMM, there are also significant differences.

Space syntax is the name for a set of analytical techniques initially developed for architecture and urban design in the 1980s (Hillier & Hanson, 1984). Using various mathematical models developed from graph theory, space syntax has been used to create social, movement and spatio-visual simulations that correlate to human phenomena (Ostwald & Dawes, 2018). Such simulations have been used to support the design of buildings and cities and to shape where people are likely to gather, move through or seek landmarks to orientate themselves (Lee & Ostwald, 2020). IMM uses data, often updated in real time or crowd-sourced, and computational simulations to incorporate human factors into transport network planning (Hsu & Wang, 2015; Lewicki et al., 2020). Whereas transport planning has traditionally been concerned with optimal route selection or congestion management, IMM adds data about human behaviours and social and cultural factors that can shape transport decision making (Pribyl et al., 2022). Applications of IMM provide nuanced and rich predictions of human navigational patterns, which in turn shape the way transportation and infrastructure operate.

Both space syntax and IMM use data to support creating or modifying buildings, cities and infrastructure. The findings developed from their use can be visualised in various ways or modified to test alternative scenarios or impacts. While both space syntax and IMM can support dynamic, cyber-physical modelling or digital twins, they tend to be used in fields where the parameters are relatively clearly stable. For example, once constructed, networks of roads are mainly static in space, and the traffic flowing on them, regardless of whether it is moving freely or blocked, falls within a predictable parameter range for IMM. Similarly, while buildings can be renovated and their spaces altered, such actions are rare in their life cycles, and for this reason, space syntax simulations tend to be most useful during the design process. However, there are examples where both space syntax and IMM have been used for less predictable situations.

Space syntax, GIS, transport modelling and network analysis have all been used to identify optimal locations for emergency facilities (Tian et al., 2023; Zhang et al., 2023). Irsyad and Hitoshi (2022), for example, used a combination of observations and space syntax to examine how people make decisions about flood evacuation. With a focus on earthquakes, Tsai and Chang (2023) analyse the effectiveness of space syntax and network analysis methods for developing mass urban evacuation strategies. Significantly, such urban-scale emergency evacuation events have been investigated using IMM. Waller et al. (2023) analyse travel disruption and traffic behaviour strategies resulting from the mass evacuation of Ukraine after the Russian invasion in 2022. They note that not only is there relatively little research available "on travel patterns during such human-driven large-scale (and sustained) disruptive events", but "studies analysing travel patterns during large-scale conflicts or invasions appear to be exceptionally limited". To respond to this knowledge gap, Waller et al. (2023) use a methodology combining crowd-sourced data and origin-destination mapping to examine patterns in travel behaviour in selected Ukrainian cities in the month after the start of the invasion. Waller et al.'s (2021) "origin-destination demand and visualisation tool Rapidex, enables the user to download and visualise road networks for any city using a capacity-based modification of OpenStreetMap". Using pervasive traffic data, they demonstrate how dynamic models of travel disruption and adaptation can be visualised.

A systematic review of applications of space syntax methods in the analysis of complex environments (Iftikhar et al., 2021) initially identified over 4000 articles, of which only nine were about stressful or perplexing spaces and activities akin to those in a disaster situation. One example of this type of research is the work of Lin et al. (2020), which used an immersive virtual reality (VR) simulation to study whether there were cultural differences in crowd behaviour during a fire emergency evacuation in metro stations. They used the Unity3D game engine to convert 3D Studio Max models of metro stations into a format suitable for the simulation. In a study with similar parameters, Wang et al. (2022) examined the impacts of lighting on peoples' wayfinding skills during an emergency evacuation. Their study also employed an immersive VR experience rendered using Unreal Engine 4, supported by a universal treadmill to enhance the sense of reality. They set up a series of experiments, which simulated an emergency fire evacuation from an underground railway station. Participants were observed, and their paths through space were traced and analysed.

Meng and Zhang (2014) compared people's experiences of exiting a VR simulation of a hotel room under two conditions. The first group exited in response to a fire alarm but without any visible signs of a fire, while the second group was confronted with a dynamic, virtual fire. By comparing the experiences of the two groups, Meng and Zhang (2014) developed rich data about the impact of a visible threat on wayfinding and behaviour. They found that the second group experienced "significantly higher skin conductivity and heart rate, experienced more stress, took longer time to notice the evacuation signs, had quicker visual search and had a longer escape time to find the exit". The simulated fire created "higher physiological and psychological stress", diminishing the capacity of this group to evacuate the burning building safely. Meng and Zhang (2014) concluded that there is a need for education and practice to improve individual and collective preparedness for disaster situations.

Given these examples of immersive experimental environments using gaming engines (Unity and Unreal) and VR, it is not surprising that "serious games" have been designed to study and prepare people for emergencies. For example, the catalyst for the research of Snopková et al. (2022) was the observation that "people tend to [follow] previously-used and known routes (to retrace) rather than follow evacuation signage. This has proven undesirable, even fatal, in emergencies and such [behaviour] calls for a better understanding of the influencing factors". In emergencies, "decisions are made according to the perceived situation prevailing, the presence of additional visible cues (e.g. fire, smoke), previous experiences, and the actions of other occupants". Snopková et al. (2022) tested a series of navigation models drawn from space syntax and network analysis, using an immersive VR simulation and an interactive environment scripted using Unity. One of their findings emphasises the capacity of this approach—VR, gaming engine and socio-spatial analysis—for engaging people in deep learning situations.

12.3 The Arts and Spatial Analytics

A recurring argument in the early 2000s was about the value of embedding the arts in a curriculum otherwise dominated by science, technology, engineering and mathematics (STEM). The acronym STEAM has come to reflect the proposition that the arts (along with the humanities and other creative disciplines) have a crucial role in STEM, developing creative insights and communicating STEM messages to the general public, thereby affecting social change. The basis for this proposition, which is repeated in many places, is that the arts "are crucial to facilitating acute and long-term insights into possible social and environmental interactions, impacts, benefits and consequences for our human condition" (de la Garza & Travis, 2018). Ludlow and Travis (2018), for example, identify the critical importance of using the arts to communicate messages about climate change and extreme weather phenomena, and Travis (2018) demonstrates how data about sociopolitical violence is ineffective for developing an understanding of its impact, without rich visualisation of its human causes and effects.

The arts engage people in galleries, libraries, archives, museums and online (Benneworth et al., 2016). As such, it is often argued that the arts have a heightened capacity to communicate with society and shape people's attitudes and behaviours (Gattenhof et al., 2021). Of equal importance, the arts provide a framework for critically examining and understanding concepts and visualising knowledge in accessible and original ways. In parallel with the increasing awareness of the power of the arts in society, there has been growing recognition that "design thinking" brings new methods to STEM, supporting divergent rather than solely convergent thinking, developing methods to promote innovation and lateral thinking (Lee et al., 2020; Yu et al., 2021). For these reasons, the idea of using arts-based thinking to explore the potential for space syntax and IMM data and methods to be used for disaster preparedness is a natural one. A practical conceptual framework for considering its potential, which appears in several scientific studies referenced previously in this chapter (Meng & Zhang, 2014; Snopková et al., 2022; Wang et al., 2022), combines immersion with the concept of rehearsal.

The dictionary definition of the verb "rehearse" describes the process of practising an activity that will later be undertaken in public. Thus, for example, a person might rehearse a piece of music and later perform it on stage. In the arts, to rehearse is not the same as to practice. The noun "practice" refers to the process of repeating various isolated techniques or skills to such an extent that they become ingrained or second nature. Thus, practice provides the foundation for rehearsal, preparing a person for a successful performance. For example, a singer might repeatedly practise scales for many years (singing ranges of notes in a precise way) but rehearse a particular aria for an opera for a few weeks before the performance. Significantly, whereas both practice and rehearsal seemingly emphasise repetition, a higher level of dynamism is required when it comes to the final performance.

The most effective performers can respond to their contexts, including audience responses, unexpected events and the vagaries of the performance space itself. At its most extreme, this type of dynamic performance takes on another characteristic, "improvisation", which refers to the capacity to extemporise or expand a performance beyond its initial content. Significantly, Kendra and Wachtendorf (2007) argue that improvisational skills are critical for assisting people to respond to unpredictable and catastrophic events, from terrorism to climate emergencies. In a similar way, Fowkes and Fowkes (2022) call for new modes of rehearsal and collaboration as a means of communicating climate emergencies to a broader world. Barker (2022) notes the role curators play in articulating the types of practice and performance needed for a world facing climate emergencies. When confronted with unpredictable events, these examples treat people's behaviours as needing practice, rehearsal and improvisation. As research cited previously in this chapter demonstrates, immersive visualisation provides a powerful means of achieving this (Lin et al., 2020; Meng & Zhang, 2014; Wang et al., 2022).

Consider Waller et al.'s (2021) visualisation tool Rapidex. It provides users with an intuitive ability to understand different traffic flows and their ecological impacts (CO_2 emissions), from the scale of the entire cities to individual neighbourhoods and buildings. The Rapidex model for Berlin (Germany), for example, offers an interactive visualisation that can be scaled and viewed in two or three dimensions (Figs. 12.1 and 12.2). Through this process, Berlin is revealed to be a city where 3.7 million inhabitants travel, on average, 6 km or 13 min of travel time per trip while experiencing a moderate level of traffic congestion (congestion index = 2.8). In contrast, Dubai (UAE), with a population of around 3.3 million inhabitants, has a lower average level of congestion (congestion index = 1.54) and an average trip length of approximately 10 km or 8 min of travel time (Figs. 12.3 and 12.4). Waller et al.'s (2021) Rapidex offers a rich tool for interactive visualisations of traffic congestion and environmental impacts across multiple scales.

Ostwald and Dawes (2013) and Ostwald (2014) provide examples of the potential for immersive simulations of space syntax data to support intuitive understanding and rehearsal. In the first of these studies, Ostwald and Dawes (2013) examine the capacity for different types of syntactic maps (space syntax network representations) to capture potential movement patterns in a famous example of Californian modern architecture, Richard Neutra's Lovell House. Using 3D simulations, along with complex spatial "dissolves", people can visualise the house first as a habitable environment and then as a complex network of movement paths and optimal places for social interaction or spatial isolation (Figs. 12.5 and 12.6). In a second work, Ostwald (2014) uses a complex mapping of the experience of movement and sound

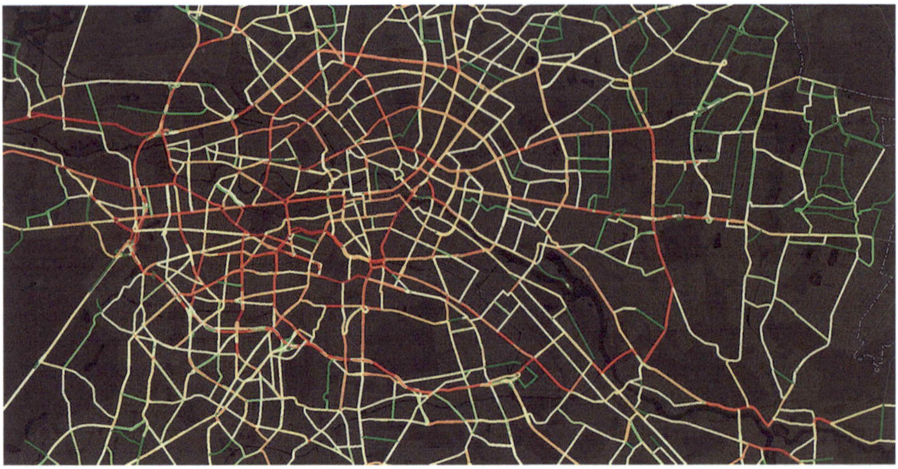

Fig. 12.1 Rapidex visualisation of Berlin urban traffic networks. (Reproduced from *TUD TMS GitHub*, https://t1p.de/3n4kd)

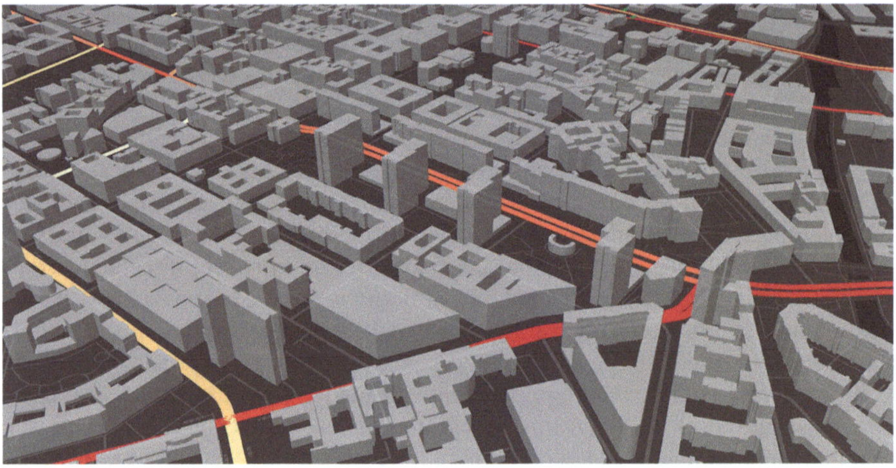

Fig. 12.2 Rapidex visualisation of Berlin buildings and neighbourhoods. (Reproduced from *TUD TMS GitHub*, https://t1p.de/3n4kd)

in another of Neutra's famous designs, the Clark House (Figs. 12.7 and 12.8). This analysis not only provides a comprehensive overview of the mathematical and spatial properties of the design, which are much debated by historians, but also offers the first immersive experience of both the house and its data. Here, people could experience the spaces as they were intended; hear simulations of sound and music, which the design was famous for; and view the mathematical maps of spatial properties and the associated mathematical data, suspended throughout the space (the coloured lines and numbers in Figs. 12.7 and 12.8).

Fig. 12.3 Rapidex visualisation of Dubai urban traffic networks. (Reproduced from *TUD TMS GitHub*, https://t1p.de/tc6r7)

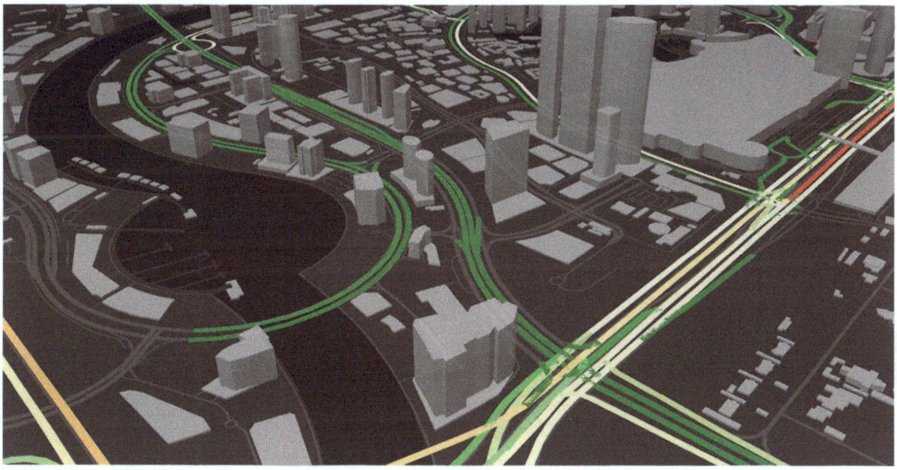

Fig. 12.4 Rapidex visualisation of Dubai buildings and neighbourhoods. (Reproduced from *TUD TMS GitHub*, https://t1p.de/tc6r7)

These examples from the authors of the present chapter demonstrate the visualisation of complex data in immersive and interactive ways, as well as support for this from AI and ML. When coupled with past research about fire and flood evacuation from buildings, along with the impacts of war on urban evacuation strategies, both space syntax and IMM have considerable potential to help people practise, rehearse and improvise when confronted with climate emergencies. What, then, is required of future research to achieve these goals?

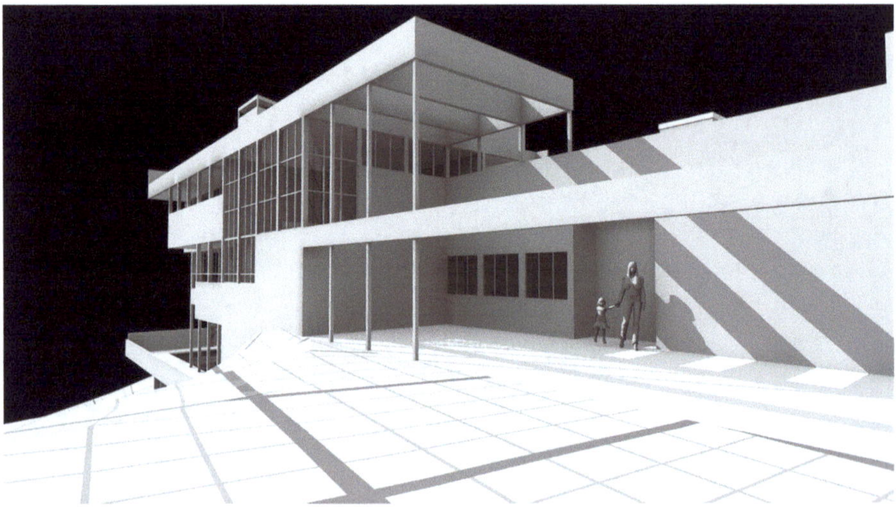

Fig. 12.5 Visualisation of Richard Neutra's 1927 Lovell House. (Ostwald & Dawes, 2013)

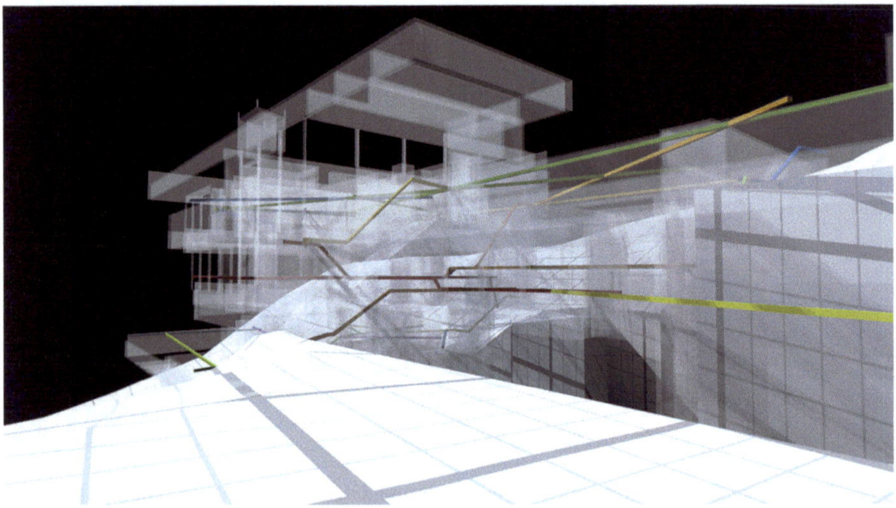

Fig. 12.6 Visualisation of one set of syntactic navigation data for the house. (Ostwald & Dawes, 2013)

12.4 Challenges and Opportunities

There are two key challenges to translating the insights of space syntax analysis into an environment suitable for supporting disaster preparedness. First, space syntax methods do not, typically at least, respond in real time to dynamically evolving environments, but this may be a necessity for practice and rehearsal in an

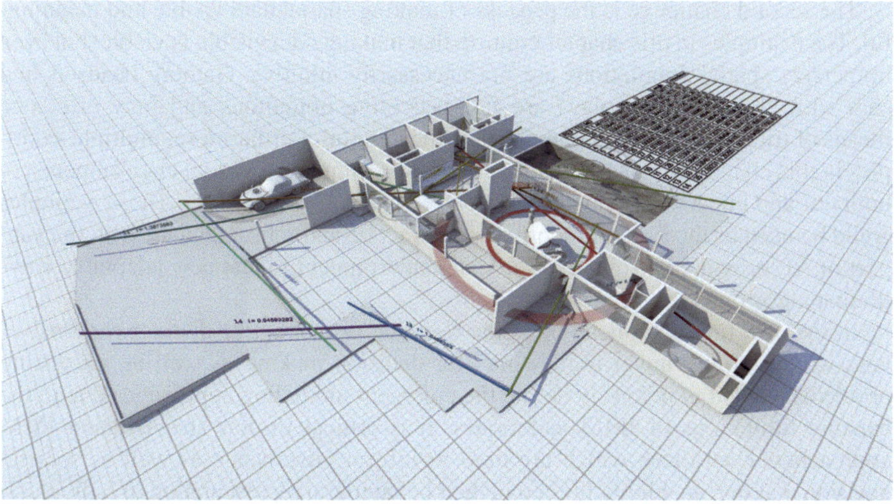

Fig. 12.7 Visualisation of Richard Neutra's 1957 Clark House, exposing syntactic navigation paths and acoustic influence maps. (Ostwald, 2014)

Fig. 12.8 Visualisation of experience of interacting with this path and map data within the house. (Ostwald, 2014)

unpredictable world. Therefore, adaptable wayfinding, without predefined paths, would be a priority for future research, wherein optimal routes to safety can evolve in response to changes in spatial hazards and altered environmental conditions. With the support of AI, socio-spatial data could be developed into an adaptive guidance system, even included in a smartphone app or the next generation of "smart glasses", to better prepare people and then guide them in real time.

The second challenge is the process of making spatial data visible and meaningful. The examples in this chapter confirm that making data visible is viable, but even immersive spatial simulations are not necessarily intuitive. Gaming headsets and cars with "heads-up" displays are fast becoming ubiquitous and may offer one means of trialling new ways to make data meaningful. Furthermore, multiple examples in this chapter use gaming engines, often in immersive virtual environments, to both test and understand people's capacity to adapt to spatial risks. These examples could be used as the basis for a process of practice and rehearsal and, with the support of AI, provide the right experimental environment to test new adaptable wayfinding algorithms.

In IMM, a future application of the research in interactive disaster responsiveness would require novel frameworks for decision making as well as for testing protocols needed to precisely define the emergent modelling approaches due to the existing limitations of IMM techniques. It is critical to note that existing IMM approaches, even those which are used to inform governance as well as societal decision making, are highly limited in their behavioural representation of individual-scale characteristics. For instance, even modern IMM approaches generally simplify highly individualised characteristics—such as the level of preparation, resilience and responsiveness—in terms of the quantification of network conditions. Moreover, most standard IMM approaches are tailored to "typical" conditions where estimating the average is the primary goal. Even for evacuation modelling and other disruption scenarios, the focus is commonly on aggregate network-level phenomena. But, by experimenting with the impact on quantifiable decision making within the highly immersive virtual environment, novel IMM solutions can be devised that are more behaviourally realistic. Insights into the individual human response to quantifiable degrees of preparation will shed light on currently unmodelled phenomena within IMM. Further, by enhancing the representation of individual decision making under varying degrees of preparation and rehearsal, the resulting functional tool would provide an entirely new policy sensitivity for IMM solutions. This increased realism enables the required decision support for policy decisions, such as potential support of societal preparedness initiatives.

12.5 Conclusion

To rehearse is to prepare for the real event. It requires both knowledge development through practice and experimentation and knowledge application in the final performance, including a capacity to adapt in real time. In this chapter, the concept of rehearsal is used to conceptualise the relationship between emergency scenarios, decision making and the built environment. Importantly, rehearsal does not occur in isolation. It requires a capacity to picture or visualise the setting, and it relies on the ability to construct a narrative and to communicate a message or emotion. While the present chapter conceptualises space syntax and IMM as supporting AI-enabled, data-driven rehearsal, both computational methods are used for visualising environments and the activities that take place in them.

Finally, while space syntax and IMM often address related problems, they each provide unique perspectives. However, further research is required to benefit from their combination within immersive virtual environments. Specifically, space syntax and IMM are currently limited in their capacity to directly address disaster scenarios and prepare society for them. Nevertheless, the use of space syntax and IMM for scenario visualisation and disaster preparedness provides a powerful new tool for (i) exploring best practices for engaging and even educating the public to prepare society for disaster scenarios and (ii) developing novel individual behavioural insights, which will improve the tools society uses to plan for these scenarios. Future research can unlock this potential and its benefits.

Acknowledgements Michael J. Ostwald's research in this chapter was supported by the Australian Research Council (FT0991309, DP1092679, DP1094154, DP220101598). The computer modelling and analysis for Figs. 12.5–12.8 were undertaken in collaboration with Michael J. Dawes.

References

Barker, J. (2022). *Museums, art and inclusion in a climate emergency*. Routledge.

Benneworth, P., Gulbrandsen, M., & Hazelkorn, E. (2016). *The impact and future of arts and humanities research*. Palgrave Macmillan.

de la Garza, A., & Travis, C. (2018). Introduction. In A. de la Garza & C. Travis (Eds.), *The STEAM revolution: Transdisciplinary approaches to science, technology, engineering, arts, humanities and mathematics* (pp. 1–12). Springer.

Fowkes, M., & Fowkes, R. (2022). *Art and climate change*. Thames & Hudson.

Gattenhof, S., Hancox, D., Klaebe, H., & Mackay, S. (2021). *The social impact of creative arts in Australian communities*. Springer.

Hillier, B., & Hanson, J. (1984). *The social logic of space*. Cambridge University Press.

Hsu, C.-H., & Wang, S. (2015). Modeling and simulation for intelligent mobility. *Journal of Computational Science, 11*, 175–176.

Iftikhar, H., Shah, P., & Luximon, Y. (2021). Human wayfinding behaviour and metrics in complex environments: A systematic literature review. *Architectural Science Review, 64*(5), 452–463.

Irsyad, H. A. W., & Hitoshi, N. (2022). Flood disaster evacuation route choice in Indonesian urban riverbank kampong: Exploring the role of individual characteristics, path risk elements, and path network configuration. *International Journal of Disaster Risk Reduction, 81*, 103275.

Jalalian, A., Chalup, S. K., & Ostwald, M. J. (2011). Architectural evaluation of simulated pedestrian spatial behaviour. *Architectural Science Review, 54*(2), 132–140.

Kendra, J., & Wachtendorf, T. (2007). Improvisation, creativity, and the art of emergency management. In H. Durmaz, B. Sevinc, A. S. Yayla, & S. Ekici (Eds.), *Understanding and responding to terrorism: NATO security through science series E: Human and social dynamics* (pp. 324–335). IOS.

Lee, J. H., & Ostwald, M. J. (2020). *Grammatical and syntactical approaches in architecture: Emerging research and opportunities*. IGI Global.

Lee, J. H., Ostwald, M. J., & Gu, N. (2020). *Design thinking: Creativity, collaboration and culture*. Springer.

Lewicki, W., Stankiewicz, B., & Olejarz-Wahba, A. A. (2020). The role of intelligent transport systems in the development of the idea of Smart City. In G. Sierpiński (Ed.), *Smart and green solutions for transport systems* (pp. 26–36). Springer.

Lin, J., Zhu, R., Li, N., & Becerik-Gerber, B. (2020). Do people follow the crowd in building emergency evacuation? A cross-cultural immersive virtual reality-based study. *Advanced Engineering Informatics, 43*, 101040.

Ludlow, F., & Travis, C. (2018). STEAM approaches to climate change, extreme weather and social-political conflict. In A. de la Garza & C. Travis (Eds.), *The STEAM revolution: Transdisciplinary approaches to science, technology, engineering, arts, humanities and mathematics* (pp. 33–66). Springer.

Meng, F., & Zhang, W. (2014). Way-finding during a fire emergency: An experimental study in a virtual environment. *Ergonomics, 57*(6), 816–827.

Ostwald, M. J. (2014). Space and sound: Harmonies of modernism and music in Richard Neutra's Clark House. *Centre, 18*, 64–77.

Ostwald, M. J., & Dawes, M. J. (2013). Differentiating between line and point maps using spatial experience: Considering Richard Neutra's Lovell House. *Nexus Network Journal, 15*(1), 63–81.

Ostwald, M. J., & Dawes, M. J. (2018). *The mathematics of the modernist villa: Architectural analysis using Space Syntax and Isovists*. Springer.

Pribyl, O., Svitek, M., & Rothkrantz, L. (2022). Intelligent mobility in Smart Cities. *Applied Sciences, 12*, 3440.

Snopková, D., Ugwitz, P., Stachoň, Z., Hladík, J., et al. (2022). Retracing evacuation strategy: A VR game-based investigation into the influence of building's spatial configuration in an emergency. *Spatial Cognition and Computation, 22*(1–2), 30–50.

Tian, F., Lei, J., Zheng, X., & Yin, Y. (2023). Integrating space syntax and location-allocation model for fire station location planning in a China mega city. *Fire, 6*, 64.

Travis, C. (2018). Spatial humanities GIS: The city as a literary, historical, and cultural STEAM lifeworld laboratory. In A. de la Garza & C. Travis (Eds.), *The STEAM revolution: Transdisciplinary approaches to science, technology, engineering, arts, humanities and mathematics* (pp. 13–32). Springer.

Tsai, M.-T., & Chang, H.-W. (2023). Contribution of accessibility to urban resilience and evacuation planning using spatial analysis. *International Journal of Environmental Research and Public Health, 20*, 2913.

Waller, S. T., Chand, S., Zlojutro, A., Nair, D., et al. (2021). Rapidex: A novel tool to estimate origin–destination trips using pervasive traffic data. *Sustainability, 13*, 11171.

Waller, S. T., Qurashi, M., Sotnikova, A., Karva, L., & Chand, S. (2023). Analyzing and modelling network travel patterns during the Ukraine invasion using crowd-sourced pervasive traffic data. *Transportation Research Record, 2677*(10), 491–507.

Wang, D., Liang, S., Chen, B., & Wu, C. (2022). Investigation on the impacts of natural lighting on occupants' wayfinding behavior during emergency evacuation in underground space. *Energy and Buildings, 255*, 111613.

Wong, A. S. W., Chalup, S. K., Bhatia, S., Jalalian, A., et al. (2012). Visual gaze analysis of robotic pedestrians moving in urban space. *Architectural Science Review, 55*(3), 213–223.

Yu, R., Gu, N., & Ostwald, M. J. (2021). *Computational design: Technology, cognition and environments*. CRC.

Zhang, L., Tian, F., Zheng, X., & Sun, Z. (2023). Spatial configuration of fire protection for historical streets in China using space syntax. *Journal of Cultural Heritage, 59*, 140–150.

Part IV
Communicating

Chapter 13
Culture, Creativity, and Climate: A Dangerous Gap in Policies of Preparedness

Stuart Cunningham, Sora Park, Kerry McCallum, Dennis Del Favero, and Janet Fulton

Abstract This chapter pinpoints a dangerous gap in official policy responses to recent fire and flood emergencies in Australia. The gap is twofold: a relative lack of focus on *preparedness* and an almost complete lack of focus on the role that *arts*, *culture*, and *creativity* can play in dealing with climate emergencies. A thematic analysis of recommendations in six major bushfire and flood inquiries commissioned between 2020 and 2022 reveals that preparedness is less a focus than resilience, recovery, and response. When preparedness was noted, it was mainly focused on government and government bodies rather than on individuals' or communities' preparedness. Several arts and culture organisations' submissions to these inquiries had virtually no representation in the recommendations. These bodies, though, tend to focus on resilience and recovery. There is an emerging academic literature supporting the preparedness perspective, but to achieve a step change in preparedness to address the accelerating climate crisis, we need coordinated use cases of fused arts and advanced technology presented elsewhere in this book. Without the ability to imaginatively preview what near-future climate shocks could *look and feel like*, it is almost impossible to believe their likelihood, let alone prepare, especially in frontline communities.

Keywords Community · Creativity · Extreme events · Government inquiries · Preparedness

---

S. Cunningham (✉) · S. Park · K. McCallum
News & Media Research Centre, University of Canberra, Bruce, ACT, Australia
e-mail: stuart.cunningham@canberra.edu.au; sora.park@canberra.edu.au; kerry.mccallum@canberra.edu.au

D. Del Favero
iCinema Centre for Interactive Cinema Research, The University of New South Wales, Kensington, NSW, Australia
e-mail: d.delfavero@unsw.edu.au

J. Fulton
College of Design & Social Context, RMIT University, Melbourne, VIC, Australia
e-mail: janet.fulton@canberra.edu.au

D. Del Favero et al. (eds.), *Climate Disaster Preparedness*, Arts, Research, Innovation and Society, https://doi.org/10.1007/978-3-031-56114-6_13

13.1 Introduction

When the Royal Commission into National Natural Disaster Arrangements was announced in 2020, after the catastrophic bushfire events in Australia in 2019–2020, then Australian prime minister Scott Morrison said: "we need to look at what actions should be taken to enhance our preparedness, resilience and recovery through the actions of all levels of government and the community, for the environment we are living in" (2020). One of the three priority areas announced in the same media release was "Improving Australia's preparedness, resilience, and response to natural disasters, across all levels of government" (Morrison, 2020). *Preparedness* was a key inclusion in the terms of reference, and the prime minister's initial announcement included *community* preparedness as part of the overall brief.

However, our analysis of the recommendations from this inquiry, and five other bushfire and flood inquiries commissioned between 2020 and 2022, revealed that *preparedness* has been less of a focus than resilience, recovery and response. When *preparedness* was noted, it was in large part focused on government (federal, state, and local) and government bodies (e.g., emergency services and government agencies) with relatively less focus on better preparation of ordinary citizens and communities. While community preparedness is specifically mentioned in the terms of reference of several of these inquiries, it remains aspirational, largely unexplored.

Maguire and Hagan claim that "a creative community may learn from the experience and teach its members how to better prepare for future disasters…so that higher levels of post-disaster resilience are attained" (2007). Yet, to a remarkable extent, art, culture and creativity were largely absent from the inquiries' recommendations. While creative practice can and does play an important part in a community's recovery process, it can also play a key role in preparedness.

This chapter is based on an analysis of inquiries responding to a series of major catastrophic events—bushfires in the Australian Capital Territory (ACT), New South Wales (NSW) and Victoria in 2019–2020, and the 2022 floods in NSW—in addition to the Royal Commission into National Natural Disaster Arrangements reporting at a national level. Along with heatwaves and cyclones, flood and fire are the deadliest hazards in Australia (Anderson-Berry et al., 2018). According to consultancy firm Deloitte (2017), the average cost over the ten years from 2007 to 2016 was $18.2 billion per year. An update in 2021 stated that natural disasters cost $35 billion per year with an expected increase to $73 billion per year by 2060. Deloitte (2021) claims that 50% of those costs are likely to be caused by bushfires and 31% by floods.

Australia's approach to emergency management comprises four phases—"prevention/mitigation, preparation, response and recovery" (Jenkins, 2015). Our analysis of inquiry recommendations showed a clear focus on response and recovery. Research has shown that Australian governments have a reactive approach to preparedness (de Vet et al., 2019) and national "resilience strategies in Australia (and elsewhere) heavily skew federal, state, and territory disaster spending towards [response] and recovery, rather than mitigation measures taken in advance of disasters to decrease or eliminate social and environmental impacts" (de Vet et al., 2019). De Vet et al. argue that Australia focuses on post-disaster in its spending priority and pre-disaster funding is limited with preparedness for households primarily focused on insurance—what they call "the politics of insurability" (2019).

Dwyer notes how natural disasters challenge government and industry organisations but stresses that communities are deeply affected and that "[g]overnment, voluntary and private organizations and communities all play a vital role in the [prevention] of, response to and recovery from bushfire" (2022). However, as argued by Cole et al. (2018), there is little of substance in inquiries that targets communities, and, as Dwyer also notes, research into public inquiries has shown that "there is little new that emerges from their deliberation" (2022) with similar findings and recommendations occurring across multiple inquiries.

The following includes a literature review of scholarly research that has examined recent inquiries in relation to fire and flood events and those that used recommendations as the primary source of data. We then review the relatively sparse literature on the role of art and creativity in relation to extreme events, followed by a discussion on how the recommendations in the reports we assessed deal with preparedness, community, arts, culture and creativity. First, though, we explain the methodology and research design we employed.

13.2 Methodology and Research Design

In an initial search, we identified inquiries carried out in the last 15 years but then chose to focus on the most recent ones, given that the extent and severity of the fire emergencies of 2019–2020 are generally accepted as having heralded a new quality of climate-induced natural disaster. Of these inquiries, we chose those that included recommendations.

Recommendations from these were analyzed (abbreviations in brackets are reflective of reference in the following discussion):

- Royal Commission into National Natural Disaster Arrangements Report: established February 2020, report released October 28, 2020 (ROYAL COMM).
- ACT Emergency Services Agency Operational Review of the Bushfire Season 2019–2020: commissioned May 2020, report released August 20, 2020 (ACT FIRE).
- Final Report of the NSW Bushfire Inquiry: established January 2020, report released July 31, 2020 (NSW FIRE).
- Inquiry into the 2019–2020 Victorian Fire Season—Phase 1 report: commissioned January 2020, report released July 31, 2020 (VIC FIRE 1).
- Inquiry into the 2019–2020 Victorian Fire Season—Phase 2 report: commissioned January 2020, report released July 30, 2021 (VIC FIRE 2).
- NSW 2022 Flood Inquiry full report: established March 2022, report released July 29, 2022 (NSW FLOOD).

The documents were imported into NVivo, and thematic analysis was employed as the analytical framework. A thematic approach enables a researcher to build "a comprehensive, contextualized, and integrated understanding" (Bazeley, 2013) of key themes in the data. While the analysis was designed to allow themes to emerge from the coding, we also included words that are key to our analytical focus, including *response*, *recovery* and *preparedness* as well as *individual*, *community/communities*, *culture*, *art* and *creative/creativity*. A further word search was carried out of the entire documents, not just the recommendations, for *art*, *culture* and *creativity* to see if there were any references in the reports themselves (outside of the recommendations) to any of those terms.

13.3 Literature Review

13.3.1 Inquiries and Recommendations

Eburn and Dovers' overview of official Australian inquiries noted that since "1939, there have been over 30 inquiries into wildfires and wildfire management and at least another 14 into floods, storms, other natural hazards, and emergency [management] arrangements" (2015). Since then, natural disasters have continued and so have inquiries. The scholarly literature analysing disaster inquiries has examined how to enhance resilience (Goode et al., 2012), how effective inquiries are (Eburn & Dovers, 2015), common themes in recommendations (Cole et al., 2018), how recommendations influence policy (Mintrom et al., 2020) and how to make sense of catastrophic events with formal inquiries (Dwyer, 2022; Dwyer & Hardy, 2016; Dwyer et al., 2021). Other research has focused on missed opportunities from inquires (McGowan, 2012), insurance as a key area recommended for preparedness (de Vet et al., 2019) as well as the development of Australia's hazard warning system (Anderson-Berry et al., 2018), which, according to Anderson-Berry et al., represents "a transition from a crisis response model to one of community preparedness, disaster mitigation and more recently, disaster risk reduction" (2018).

Cole et al. (2018) contend that managing to achieve better outcomes is a key aspect of reviews and inquiries into disasters and emergencies in Australia. Reviews and inquiries assist the emergency management sector, the government and individuals to prepare, respond to and recover from disasters and have a common objective: "to identify the cause and consequences of disasters and recommend future practices for better outcomes. In some cases, they attribute responsibility or blame for failings" (Cole et al., 2018). Cole et al. (2017, 2018) conducted research into whether recommendations from disaster inquiries were valuable for Australian emergency services and useful as a national information resource. They examined 1336 recommendations from 55 major event inquiries between 2009 and 2017 to discover if there were recurrent themes and found 32 common themes in the recommendations, ranging from better coordination between agencies to better community warnings and communication, to government's role, funding, volunteers and personal responsibility (see Cole et al., 2018 for a full list). *Community education and preparedness* was a recognised theme and included in 25 of the examined inquiries with 58 recommendations overall, as was *pre-fire season preparation* in 16 inquiries (with 30 recommendations). Cole et al. also note the theme *personal responsibility*, which covers the emerging policy of shared responsibility, but found that there were few recommendations that targeted communities and community members. It is also telling that there is no mention of creativity, culture or creative arts either in preparing for catastrophic events, which is our area of interest, or in response and recovery.

In 2012, McGowan claimed that there was a "serious gap in disaster management policy in Australia" (2012) because the focus was on response and recovery and not across the full spectrum of prevention, preparedness, response and recovery. McGowan further claimed that there is a disproportionate allocation of resources to response and recovery with prevention and preparation "badly neglected" (2012). This claim is reiterated by de Vet et al., who state that only "3% [...] of disaster spending [in Australia] goes towards mitigation" and "[m]ost governments prioritize disaster response and recovery over risk reduction and mitigation" (2019). Despite our 2019–2022 corpus of inquiries including *preparation* in their terms of reference, clearly the primary focus is on response and recovery in their recommendations. Furthermore, regardless of the increasing understanding of the value of arts, culture and creativity in enabling recovery from catastrophic events, these concepts are rarely included in relation to preparing a community.

13.3.2 Arts, Culture and Creativity

There are several organisations whose work underpins the role that arts, culture and creativity can play, and currently plays, in preparing for, dealing with and recovering from extreme flood and fire events in Australia. For example, Creative Recovery Network (CRN) are dedicated to "collaborate with the creative sector—individual artists, arts workers, and arts organisations—to support the ongoing activation of creative programs in disaster preparedness, response and recovery" (2023). Local Government NSW (LGNSW) have collaborated with CRN and others to form a Creative Recovery Taskforce to "provide recommendations on how the creative arts can assist with disaster management" (LGNSW, n.d.) through all phases of disasters, including preparedness. The Creative Recovery Taskforce's fourth recommendation specifically states:

> Formally recognise creative practitioners and culture and the arts, as an essential component of Australia's disaster management capacity and a key component of the nation's preparedness, recovery and resilience capability, delivering trauma informed practice and helping people prepare for and cope with disasters. (LGNSW n.d.)

Other organisations in Australia that have a keen interest in how arts and creativity can play in extreme fire and flood events include Regional Arts Australia, Creative

First Aid Alliance, Australian Museums and Galleries Association, Firesticks Alliance Indigenous Corporation, Flood Diaries Citizen Storytelling Project, GLAM Peak and the National Association for the Visual Arts. All these bodies made submissions to some of the 2019–2022 inquiries on which we are concentrating.[1] These very valuable bodies, though, like their inquiry submissions, tend to focus on resilience and recovery. There is a relative lack of focus on preparation, especially at the grassroots community level. When the focus is on preparation, it is often with respect to local cultural infrastructures and assets rather than broad-based cross-sector and community readiness.

Internationally, the Federal Emergency Management Agency (FEMA) in the United States acknowledges the role that emergency workers and managers take in preparing communities for disasters, but they also acknowledge the important role that artists can play: "[It continues with] exploring how we can partner with artists to communicate risk and building a culture of preparedness across the nation" (2020). FEMA claims that, for example, community art can generate meaningful context when educating about and communicating risk. Other forms noted by FEMA include music, performance art, theater and placemaking. West, Balog-Way and Phillips, in a blog on the World Bank Blogs site, note:

> Art can inspire people to think about disaster risk and resilience in ways that science, data, and numbers cannot…Poetry, painting, photography, music and performances can all tell stories across cultural barriers, building empathy for communities who are facing increased risk from hazards and climate change. Emotions evoked by art can convey a sense of urgency for preventing and preparing for disasters. (2019)

Scholarly research in this area skews toward a stronger focus on art as part of the recovery process as distinct from preparation. And art in this sense includes music, drama, poetry, creative writing, photography and visual arts such as painting. Researchers have investigated the importance of art in healing from disasters (Green, 2021; McManamey, 2009; Scarce, 2022; Smilan, 2009; Van Laar, 2022), as a tool of resilience and recovery (Garavaglia, 2019; McManamey, 2009), improving health and well-being (Green, 2021; McManamey, 2009), maintaining local knowledge and the history of a community (McManamey, 2009) and community building (Brien & Hawryluk, 2011; Garavaglia, 2019; Van Laar, 2022). The value of art therapy for children after disasters is a key research area (for a summary, see Haring et al., 2018 and Smilan, 2009). Joseph Scarce (2022), in *Art Therapy in Response to Natural Disasters, Mass Violence, and Crises*, argues that the key is "the nourishment that art-making provides for continued healing and connection to the creative flow process" (2022) and "we bring with us hope through art. We bring communities together and show that there are others who are with them in crisis and that they are not alone" (2022). While Scarce's collected chapters discuss devastating disasters and the value of art therapy in building communities, the common thread is that that value is manifest in the recovery phase.

---

[1] For examples of who made submissions to inquiries, see ROYAL COMM, VIC FIRE 1 and 2, and NSW FLOOD.

McManamey (2009) reinforces this perspective, examining how a book that documented people experiencing the 2006 fires in East Coast Tasmania played a role in the resilience and healing of community members. The book included images, stories, reports, artworks and poetry. McManamey conducted interviews with community members, volunteers, service providers and politicians about their sense of the importance of the book. Several key themes emerged, and, although the majority were concerned with response and recovery, there were two that speak to our argument: *support for future preparedness* and *attention drawn to "non-preparedness."* The first noted how "examples within the publication supported preparedness and enabled community members—empowering the community" (2009). McManamey also found that a major outcome of this community-led initiative was "*future preparedness*—ongoing motivation and interest in communities to attend further sessions on safety" (2009). While primarily focused on recovery and healing, McManamey notes how written data and visual expression can also "inform on issues pertaining to a number of areas of recovery policy and preparation for future events" (2009).

Researchers such as Haring et al. (2018) have examined preparation of children for natural disasters through education. They note how the Australian curriculum now contains disaster education but argue for an arts-based curriculum, which includes poetry, and visual and performing arts, as a more effective and age-appropriate way to prepare children.[2]

13.4 Findings

13.4.1 Preparedness

The examination of the recommendations from the six reports showed a clear imbalance between *preparedness*, *response*, and *recovery*, with *preparedness* demonstrably lower. Additional examination further found that *preparedness* recommendations focused on expectations for, and success of, preparedness by government and government bodies, with much less attention for communities or individuals. This might be explained away by the fact that the primary audience of the inquiry reports were governments, but that is a weak argument when it is considered that community preparedness and even empowerment occurs often in the general language of official reports. Three of the reports' recommendations had codes that related to *preparedness_individual* and *preparedness_community*: the Royal Commission (both), the NSW bushfire inquiry (community), and the NSW flood inquiry (individual). When *preparedness* was applied to individuals and communities, it was

[2] For a review of research into how children are educated for disasters during their schooling, see Haring et al. (2018).

primarily in the form of what the government could do rather than what the community or individual could do:

- Educating the public about evacuation, essential services, sheltering facilities, and hazard reduction.
- The government building resilience.
- The government providing comprehensive information.
- Ensuring the public understands the risks of living in certain areas.

13.4.2 Community Preparedness

The Royal Commission did, however, include a chapter on community preparedness called "Community education." This chapter was seven pages long (0.01% of the total 594-page report) and generated one recommendation out of 80:

> Recommendation 10.1 Disaster education for individuals and communities
>
> State and territory governments should continue to deliver, evaluate and improve education and engagement programs aimed at promoting disaster resilience for individuals and communities. (ROYAL COMM, 2020)

This chapter included suggestions on joint responsibility for preparedness, including understanding hazards in their immediate area, protecting themselves and understanding the recovery process as well as the need for comprehensive and ongoing government education programs:

> Education is key to informing and empowering communities. Education and engagement programs should account for changing risk profiles and community demographics to ensure that they are fit for purpose and support individual and community resilience to natural disasters. Programs must have all of the information people need to make informed decisions. (2020)

It should be noted that other reports did include references to *community,* but it was typically minor or in relation to response or recovery. Some examples include liaising with Indigenous communities (NSW FLOOD; NSW FIRE 1), use of social media during disasters (NSW FLOOD), ongoing personal and community effects (NSW FLOOD), community-led recovery (VIC FIRE 2), understanding fuel management (VIC FIRE 1), responsibilities of the community (VIC FIRE 1), building resilience in communities (NSW FIRE), community warning systems (NSW FIRE) and community volunteers (NSW FIRE).

But there is also a discernible shift in the most recent inquiry into the NSW floods:

> Preparedness is discussed in relation to emergency management and our natural and built environment. But an important component of preparedness is at a personal or family level. Failure to prepare at this level makes preparations at other levels more difficult and expensive. Targeted public and school education is required to build intergenerational knowledge and enable whole families to engage in disaster readiness. (Fuller & O'Kane, 2022)

It stresses preparedness under uncertainty, which means "Australia's research [capacity] is critical to improving our ability to imagine and predict what may happen in the future" (Fuller & O'Kane, 2022), and notes, radically, that "community was often more effective at saving community than Government" (Fuller & O'Kane, 2022) and recommends a Community First Responders Program.

13.4.3 Art(s)

Art did not show up in any of the recommendations, although it was noted several times through a keyword search of the whole document corpus. For example, VIC FIRE 1 noted how cultural burning had promoted "the conservation of all cultural sites in the landscape—for example rock art and canoe trees" (2020) as an example of managing the land or preparing. VIC FIRE 2 discussed funding "new infrastructure such as playgrounds and art murals" (2021) as positive for communities and a way to assist in recovery while also noting CRCs [Community Recovery Committees] and community groups told IGEM [Inspector-General for Emergency Management] that "funding resilience and preparedness projects was more difficult, as this work does not tick a recovery box on an application form" (2021). NSW FLOOD mentioned that creative arts was one way to rebuild economically but, again, as part of the recovery process. The ROYAL COMM included Indigenous art sites as part of Australia's heritage places, and NSW FIRE noted Indigenous rock art in the Blue Mountains as a high-value cultural asset that needs to be managed in preparation for future bushfires. The ACT FIRE report had no mention of *art*.

13.4.4 Culture

There was little mention of culture, in its narrow definition as "artistic and intellectual activities" (Williams, 1981), in the recommendations. However, if we take Raymond Williams' wider definition into account, with culture defined as "a distinct 'whole way of life'" (Williams, 1981), there are several mentions of Indigenous cultural practices, particularly instigating cultural burning, and therefore in preparing, as well as within workplaces and organisations. *Culture*, in this more holistic sense, was noted in both the recommendations and the keyword search.

Culture was a theme in relation to workplace culture (ACT FIRE, VIC FIRE 1, VIC FIRE 2), the "Australian way" of volunteering (ROYAL COMM, NSW FLOOD), the culture in emergency organisations (NSW FLOOD, ROYAL COMM, NSW FIRE) and Australian identity (ROYAL COMM). It was also noted in how cultural assets were protected (ACT FIRE) during the response phase. *Culture* was also used in a vague "coverall" statement as one of the outcomes for the Victorian government's recovery programs: "Victorians are connected to people, places and culture" (VIC FIRE 22021). In regard to preparedness, support for tourism businesses to prepare was also mentioned in the NSW FIRE recommendations.

However, it was in Indigenous cultural practices that *culture* was primarily noted through the keyword search. The VIC FIRE 2 report included a chapter, "Aboriginal culture and healing," and bluntly summed up the importance of Indigenous culture: "That past influences the experience of Aboriginal people during emergencies and requires consideration when using the concepts of Aboriginal culture and healing to underpin strategies for their recovery" (2021). While this quote may seem to be about response and recovery only, this chapter also includes a clear mandate for preparedness:

> In 2021 the Aboriginal Community Mitigation and Crisis Management Grants were made available to increase the preparedness and resilience of Aboriginal communities to emergencies. There are two streams of funding; one to support Aboriginal organisations to undertake emergency planning and mitigation initiatives on bushfire-affected Country, and the other to support projects that improve Aboriginal infrastructure in a way that increases preparedness for future emergency events. (2021)

The Royal Commission also included a chapter called "Indigenous land and fire management" that recommended engaging with traditional owners to manage the land.

Regarding other forms of Indigenous culture, recommendations in the NSW FIRE, NSW FLOOD and ROYAL COMM specifically noted the importance of engaging with traditional owners to understand land and fire management in preparedness, response and recovery. Other recommendations included engaging with local Aboriginal communities "in emergency planning and preparation" (NSW FIRE, 2020; NSW FLOOD, 2022) and to provide appropriate support during evacuations (NSW FIRE; NSW FLOOD). Other recommendations included "an Indigenous led cultural landscape restoration strategy for the Northern Rivers" (NSW FLOOD, 2022), Indigenous stewardship practices and representation within government bodies.

13.4.5 Creative/Creativity

Creative and *creativity* were mentioned once in the NSW Flood Inquiry and it was in relation to a submission where a community member discussed the difficulties of relocating from Lismore and it was to do with recovery:

> I adore Lismore and this region generally and believe that it is truly unique and I cannot imagine wanting to live anywhere else but within the broader community here. They have carried me through extremely difficult times with creativity, much laughter and incredible love & generosity and I feel that I have an important place & role here within the Northern Rivers which I do not wish to lose. (NSW FLOOD, 2022)

There were no other instances of *creative* and *creativity* in any of the other inquiry documents and none in the recommendations.

13.5 Conclusion

We conducted a thematic analysis of recommendations from recent government inquiries into catastrophic Australian bushfire and flooding events. A key aim was to discover how important *preparedness* is within these inquiries but also to find out the role that arts, culture and creativity can play in preparing communities for catastrophic events. What we found is that the inquiries were skewed heavily toward government and government agencies with considerable fewer recommendations focused on communities' own resourceful agency. With several organisations and some research acknowledging the importance of arts, culture and creativity to prepare communities, it was also notable that the recommendations had little to no mention of these concepts. However, culture, when defined as "a whole way of life" (Williams, 1981) rather than "artistic and intellectual activities" (ibid.), was recognised, particularly First Nations cultural practices.

This chapter has pointed to a dangerous gap in official policy responses to fire and flood emergencies in Australia. Given the lack of focus on *preparedness* and an almost complete lack of focus on the role that *arts*, *culture* and *creativity* can play in dealing with climate emergencies, it is clear that the case to achieve a step change in preparedness to address the accelerating climate crisis needs the kinds of coordinated approach fusing arts and advanced technology canvassed elsewhere in this book. Without the ability to imaginatively preview what near-future climate shocks could *look and feel like*, it is almost impossible to believe their likelihood let alone prepare, especially in frontline communities. This presents an opportunity for Australian creatives to partner with scientists and technologists in order to transform disaster depiction from communicating a reactive observation of the aftermath to socialising visceral interactions that sparks the imagination and optimises readiness. Thereby, the community can become an active protagonist safely rehearsing life-saving responses in real time. An approach that focuses on longer-term engagement, using immersive visualisation and associated strategies, with communities of practice ranging from experts in emergency management to first responders embedded in frontline communities, will help to address dangerous gaps in official policy.

References

Anderson-Berry, L., Achilles, T., Panchuk, S., Mackie, B., et al. (2018). Sending a message: How significant events have influenced the warnings landscape in Australia. *International Journal of Disaster Risk Reduction, 30*, 5–17.

Bazeley, P. (2013). *Qualitative data analysis: Practical strategies*. Sage.

Brien, D. L., & Hawryluk, L. (2011). Facilitating 'storied relationships to a place. *TEXT Special Issue Website Series, 12*, 1–12.

Cole, L., Dovers, S., & Eburn, M. (2017). *Major post-event inquiries and reviews: Review of recommendations*. Bushfire & Natural Hazards CRC.

Cole, L., Dovers, S., Gough, M., & Eburn, M. (2018). Can major post-event inquiries and reviews contribute to lessons management? *Australian Journal of Emergency Management, 33*(2), 34–39.

Creative Recovery Network. (2023). *Who we are*. https://creativerecovery.net.au. Accessed 17 Oct 2023.

de Vet, E., Eriksen, C., Booth, K., & French, S. (2019). An unmitigated disaster: Shifting from response and recovery to mitigation for an insurable future. *International Journal of Disaster Risk Science, 10*, 179–192.

Deloitte Access Economics. (2017). *Building resilience to natural disasters in our states and territories*. Deloitte. https://t1p.de/k5gux. Accessed 17 Oct 2023.

Deloitte Access Economics. (2021). *Special report: Update to the economic cost of natural disasters in Australia*. Deloitte. https://t1p.de/h8stz. Accessed 17 Oct 2023.

Dwyer, G. (2022). *Making sense of natural disaster*. Springer.

Dwyer, G., & Hardy, C. (2016). We have not lived long enough: Sensemaking and learning from bushfire in Australia. *Management Learning, 47*(1), 45–64.

Dwyer, G., Hardy, C., & Maguire, S. (2021). Post-inquiry sensemaking: The case of the 'black Saturday' bushfires. *Organization Studies, 42*(4), 637–661.

Eburn, M., & Dovers, S. (2015). Learning lessons from disasters: Alternatives to royal commissions and other quasi-judicial inquiries. *Australian Journal of Public Administration, 74*, 495–508.

FEMA. (2020). *Guide to expanding mitigation: Making the connection to arts and culture*. Fema. https://t1p.de/wlgov. Accessed 17 Oct 2023.

Fuller, M., & O'Kane, M. (2022). *Flood inquiry vol. 1: Summary report*. NSW Gov.

Garavaglia, A. M. (2019). *Disaster preparedness and recovery* (Unpublished Masters thesis). Drexel University.

Goode, N., Spencer, C., Archer, F., McArdle, D., et al. (2012). Have we learnt the lessons from disaster inquiries? *Injury Prevention, 1*(18), A75.

Green, D. (2021). Enduring liminality: Creative arts therapy when nature disrupts. *Creative Arts in Education and Therapy, 7*(1), 71–91.

Haring, U., Sorin, R., & Caltabiano, N. (2018). Circling the cyclone: Children's understanding of natural disasters through the arts. *International Journal of Pedagogy & Curriculum, 25*(4), 1–15.

Jenkins, S. (2015). Resilience: The new paradigm in disaster management—An Australian perspective. *World J. Engineering & Technology, 3*, 129–139.

LGNSW. (n.d.). Creative arts during disaster preparedness and recovery [media release]. *Local Government NSW*. https://t1p.de/od4lj. Accessed 17 Oct 2023.

Maguire, B., & Hagan, P. (2007). Disasters and communities: Understanding social resilience. *Australian Journal of Emergency Management, 22*(2), 16–20.

McGowan, J. (2012). A missed opportunity to promote community resilience?—The Queensland floods commission of inquiry. *Australian Journal of Public Administration, 71*(3), 355–363.

McManamey, R. M. (2009). Arts, health, community resilience and healing: The importance of the story. In *Proc. 10th Nat. Rural health Conf.* National Rural Health Alliance.

Mintrom, M., O'Neill, D., & O'Connor, R. (2020). Royal commissions and policy influence. *Australian Journal of Public Administration, 80*, 80–96.

Morrison, S. (2020). National Royal Commission into Black Summer bushfires established *Australian Government*. https://t1p.de/rdyxh. Accessed 17 Dec 2023.

Scarce, J. (2022). *Art therapy in response to natural disasters, mass violence, and crises*. Jessica Kingsley.

Smilan, C. (2009). Building resiliency to childhood trauma through arts-based learning. *Childhood Education, 856*, 380–384.

Van Laar, C. (2022). Art therapy first aid: Growing capacity with art therapists in communities affected by Australian bushfires. In J. Scarce (Ed.), *Art therapy in response to natural disasters, mass violence, and crises* (pp. 134–147). Jessica Kingsley.

West, J. M., Balog-Way, S., & Phillips, E. (2019). *Harnessing creativity for change: The art of resilience*. World Bank Blogs. https://t1p.de/lzg1h. Accessed 17 Oct 2023.

Williams, R. (1981). *Culture*. Fontana.

Inquiries

ACT FIRE. (2020). *ACT Emergency Services Agency operational review of the bushfire season 2019–20.* https://t1p.de/z2wvc. Accessed 17 Oct 2023.

NSW FIRE. (2020). *Final report of the NSW Bushfire Inquiry.* https://t1p.de/tj0x3. Accessed 17 Oct 2023.

NSW FLOOD. (2022). *NSW 2022 flood inquiry full report.* NSW Independent Flood Inquiry, Sydney. https://t1p.de/w1pq2. Accessed 17 Oct 2023.

ROYAL COMM. (2020). *Royal Commission into National Natural Disaster Arrangements report.* https://t1p.de/ftui7. Accessed 17 Oct 2023.

VIC FIRE 1. (2020). *Inquiry into the 2019–20 Victorian fire season—Phase 1.* https://t1p.de/cvhtj. Accessed 17 Oct 2023.

VIC FIRE 2. (2021). *Inquiry into the 2019–20 Victorian fire season—Phase 2.* https://t1p.de/hc1ux. Accessed 17 Oct 2023.

Chapter 14
Creatively Reimagining Place and Community in a World of Extreme Weather

Helena Grehan, Belinda Smaill, and Michael J. Ostwald

Abstract This chapter explores connections between place, community and narrative in the context of a world beset by extreme weather events. Drawing on insights and readings from three disciplines—theatre studies, screen studies and architecture—the chapter constructs a rich picture of the ways these fields contribute to definitions of place and can potentially enhance disaster preparedness and recovery activities. Edward S. Casey's theories of place and community provide a connecting thread throughout the chapter, along with his ideas about selfhood, "implacement" and the environment as a source of danger. As both an example of a work that begins to address these themes and a catalyst for discussion, the chapter examines the television series *Fires* (Ayres et al., Fires [TV Series]. Australian Broadcasting Corporation, 2021), which dramatises the 2019–2020 Black Summer fire season in Australia. Starting with a broad view of the context depicted in this series, the focus then shifts to individual experience and finally emotional responses. The chapter concludes by considering future research opportunities through which the disciplines of theatre studies, screen studies and architecture can leverage applications of advanced technology to contribute to disaster preparedness, responsiveness and recovery.

Keywords Architecture · Community · Digital arts · Extreme events · Media arts · Place · Storytelling · Theatre studies

---

H. Grehan (✉)
Western Australian Academy of Performing Arts, Edith Cowan University, Mount Lawley, WA, Australia
e-mail: h.grehan@ecu.edu.au

B. Smaill
School of Media, Film & Journalism, Monash University, Caulfield East, VIC, Australia
e-mail: belinda.smaill@monash.edu

M. J. Ostwald
School of Built Environment, The University of New South Wales, Kensington, NSW, Australia
e-mail: m.ostwald@unsw.edu.au

D. Del Favero et al. (eds.), *Climate Disaster Preparedness*, Arts, Research, Innovation and Society, https://doi.org/10.1007/978-3-031-56114-6_14

14.1 Introduction

This chapter argues for the importance of a deep understanding of place and community for thinking anew about engagement with extreme weather events. It benefits from the authors' location in three distinct disciplines: theatre studies, screen studies and architecture. We place these in dialogue to explore ideas of place and community as they are imagined creatively through these combined disciplinary approaches. The Australian television drama series *Fires* (Ayres et al., 2021) was produced in the wake of the 2019–2020 Black Summer fire season, which led to the loss of 34 lives, destroyed over 3000 buildings and burnt more than "12 million hectares of forests and agricultural areas across southeastern Australia" (Lindenmayer & Taylor, 2020). A fictionalised account of this fire season, *Fires* offers an anchor to this analysis, allowing us to deploy the different disciplinary approaches in concert to offer a multilayered understanding of what place and community might mean in the context of extreme events. In turn, we are interested in how this multidisciplinary investigation might highlight new pathways for understanding creative practice as an instrument for disaster preparedness. The capacity of advanced technology—virtual reality (VR), digital interactive narrative and artificial intelligence (AI)—to support the creation of immersive even visceral experiences is central to several of these pathways.

Our discussion of "place" and "community" references different scholarly trajectories but finds a conceptual anchor in the work of Edward S. Casey, who argues that "to be at all—to exist in any way—is to be somewhere, and to be somewhere is to be in some kind of place. Place is as requisite as the air we breathe, the ground on which we stand, the bodies we have. We are surrounded by places" (1998). While place is everywhere, it is not so ubiquitous as to be meaningless. The places we inhabit are meaningful because they are inscribed within our lives, experiences and connections. Self and place are inseparable. We also acknowledge that places are redolent with the mutual reliance on all forms of life and their habitat. In this respect, we can consider Casey's point that "we are not masters of place, but prey to it" (2001) and redeploy it to highlight human vulnerability to and reliance upon the nonhuman environment and its dangers.

For places to retain meaning, we must consider how it is that we inhabit them. It is here that the notion of community comes in. We form communities in place. In co-location, we take on shared responsibilities for our places, whether they are local communities, interest groups, shared work environments, sporting groups, schools or any other place where bodies and subjectivities come together. As Casey points out, "the very word 'society' stems from socius, signifying 'sharing'—and sharing is [done] in a common place" (1998). He goes on to explain that "as Victor Turner has emphasized, a communitas is not just a matter of banding together but of [bonding] together through rituals that actively communalize people—and that require [particular] places in which to be enacted" (ibid.). We are interested in how these communities and subjects are connected by the storytelling potential of architecture, performance and nonfiction screen culture.

Both a sense of place and a sense of community are often described in terms of their contexts or settings, which commonly encapsulate landscapes or buildings. Indeed, the close connection between a landscape (its climate, topography and flora) and the architecture that arises organically within it (and responds to its opportunities or threats) is a central concept in the theory of genius loci. This theory links the phenomenology of a place with the community that lives and interacts in it and the architecture that shelters or sustains this community (Norberg-Schulz, 1980). This aligns with Casey's notion of habitus, a concept that ties self and place together. Subjects actively relate to places by way of "habitation" as social relations inform and shape our relation to places including the built environment. He argues that "in any journey through the place-world, we live out our bodily habitudes in [relation] to the changing spatiality of the scenes we [successively] encounter" (2001). Viewed through this lens, architecture and landscape are integral to both a sense of place and the characterisation of a community. Furthermore, places are not static; they can be created. This idea, known as "place-making", was first developed in the 1960s, coming to prominence in the work of Jane Jacobs (1961) and William H. Whyte (1980). Place-making connects space, architecture and public arts by way of the formation of social and cultural groups or support for active storytelling.

Audiovisual forms, such as film and television, reconceptualise a sense of place for the viewer. They can enable an experience of place as being distant, removed from the location and experience of the viewer with landscapes and human activities, to be consumed from afar without significant engagement (Cowie, 2011). Filmmakers and producers, however, often seek a meaningful engagement with place. This is particularly the case with documentary filmmakers, who frequently grapple with the possibilities for expressing place in ways that enhance its meaning in purposeful ways. If Casey tells us that "to be somewhere is to be in some kind of place", and this place is crucially tied to selfhood, the place in which viewers are asked to encounter individuals on screen conveys explicit or implicit information about the lives, experience and, indeed, habitus of the subject at the centre of the storytelling. This rests on the way editing, framing, sound design and camerawork combine to create strategies of realism and organise perception. These strategies both produce audiovisual evidence and direct our attention, whether affective or cognitive. A sense of place can combine with a sense of community to, for example, enhance a feeling of belonging and home (e.g. filmmaking in occupied territories (Cowie, 2011)); aid in working through collective and individual experiences of trauma by examining the detail of places and what occurred there (e.g. wartime atrocities or family violence (Walker, 2005)); make sense of possible future scenarios of environmental change (Salazar, 2015); or represent the complexity of places as landscapes, ecologies and social relations in ways that would not be perceptible without mediation (Smaill, 2016). Finally, the production of the documentary itself can be a collaborative practice that filmmakers can use to create communities by facilitating participation in the creative process of film production (Miller et al., 2017).

Theatre and performance studies take myriad approaches to considering, rendering and negotiating a sense of place and community, both within the performance space and out into the lived reality of spectators and participants. From realist depictions of actual places and communities in mainstream practice to the avant-garde imaginings of experimental theatre and performance, these concepts are continually being negotiated and remain in flux. What is significant, in this context, for a consideration of both place and community is the fact that these concepts are engendered or imagined in an environment that is both ephemeral and proximate. Indeed, the intimacy of the performance space allows places and communities to feel inhabited, lived in and close. Even in the most abstract of work, the conditions of the theatrical encounter in a live (or primarily live) space with actors, sets, soundscapes and scripts create a sense of intimacy that means the representation of place and community becomes heightened or vivid. This is the case across all forms of theatre and performance, and it is something that occurs even when the work itself is attempting to unsettle, disrupt or antagonise its audience—or, indeed, these notions. Casey's idea of "implacement" is helpful in this context. As outlined in the introduction to this chapter, he urges a consideration of place in the here and now. He explains that "place itself is concrete and at one with action and thought" (2009). This idea of considering place and, as he further explains, "your immediate placement or 'implacement'…counts for much more than is usually imagined" (ibid.). Drawing on this, we argue that within the live, embodied locale of the performance or theatrical space, this idea of implacement can become heightened, and, as a result, it can generate a sense or experience in which spectators think deeply about their sense of place. With that then naturally come considerations of community and belonging.

14.2 Building Place: Architecture, Theatre and Film

Encounters with the built environment, the intimacy of performance and the moving image occur in space, but they also engage an embodied relationship with what we are referring to as a more dynamic and culturally infused notion of place. We suggest that an engagement with creative practice and the imaginative potential for responding to extreme events must be connected with the rich possibilities of storytelling. Narrative, or storytelling, takes each of our disciplinary approaches into the domain of networks that connect space and time or cause and effect. From the perspective of architecture, buildings reflect the values or aspirations of the society that created them and the climate they provide shelter from. A Swiss chateau, for example, uses local stone at its base for stability and timber sourced from nearby trees for upper-level floors and walls, and has a steeply pitched roof to shed heavy winter snow. In contrast, an Australian historic homestead is typically a low-lying, pitched roofed structure, which sits on stumps, just above the ground. It has wide verandas to provide shade and capture the breeze, and its walls are lined with local timber. A corrugated metal roof allows for rainwater to be captured in tanks, and windmills pump water from nearby dams. Significantly, both the chateau and the homestead reflect the surrounding landscapes, the former situated among steep, snow-capped mountains and the latter in a wide, flat expanse of the Australian bush. These buildings each tell a story about the local environment, the materials available and the people who built them. This is one type of narrative that buildings communicate.

Another type is associated with architecture's role in depictions of place or community. For example, the tall, narrow canal houses of Amsterdam provide the real-life setting for *The Diary of Anne Frank* and the fictional counterpart for Albert Camus's novel *The Fall.* These canal houses are also a critical place-making device in the virtual reconstruction of Amsterdam depicted in the computer game *Call of Duty: Modern Warfare II*. New York's Dakota building, on the Upper West Side, is a place of isolation and otherness, separate from the world that surrounds it, in the film *Rosemary's Baby* and in Jack Finney's novel *Time and Again.* A New York apartment is also the setting for *Façade* (2005), by Michael Mateas and Andrew Stern. This AI-powered interactive digital narrative is set in a bleak modern apartment, with the only outlook an ominous city skyline. The sense of isolation created in *Façade* complements the narrative, which has the reader/player observing the relationship breakdown of two friends. The domestic dispute is not scripted; it evolves in dynamic and unpredictable ways, depending on the reader's/player's interactions and responses.

As interactive digital narratives, coupled with AI-supported virtual models of the world, become more common, the use of architecture to evoke a sense of place has become even more important. For example, the computer game *Tom Clancy's The Division* (Ubisoft, 2016) features a post-apocalyptic vision of the neighbourhood around the Dakota building, as well as parts of Greenwich Village, a location made famous by Jane Jacobs' (1961) social critique, *The Death and Life of Great American Cities*. While, for Jacobs, the federal-style hipped roofs and brownstone-clad buildings of Greenwich Village provided a positive sense of place and community, in *The Division*, their ruined state is a symbol of the breakdown of a liberal, democratic society. In these examples from film, gaming, media, fiction and nonfiction, architecture is a core element of a larger narrative, enclosing, protecting and framing society.

From its birth in Athens, theatre has always been interested in and concerned with tragedy, as a key storytelling mode. Tragedy provides narrative and dramatic energy for much contemporary theatre and performance practice. There is often a tension at play between the desire to tell big, dramatic and difficult stories and the desire not to reduce or simplify these. Not to make things easy or didactic, and as such to close over the possible meanings an audience might glean from the experience. Unlike other modes of representation, there is no formula that assists with the framing of a narrative, especially in contemporary performance works. Instead, there is a negotiation that takes place between this desire to tell or show and at the same time to hide or obfuscate. Much of the best work dwells in the murky in-between. In abstract imagery and text, in partial stories and fragments, in intertextual allusion. In the capacity to reveal glimpses of an idea and at the same time to eschew saying or telling in any didactic or literal way. Making sense of the work is left to the audience, to their interest in and capacity for imagination and meaning making. Some works that reflect this play between telling and hiding include Caryl

Churchill's *Escaped Alone* (2016), in which four women in their 70s discuss life, death, murder and the banal. As Susannah Clapp notes, "Caryl Churchill's [magnificent] new play unleashes an intricate, elliptical, acutely female view of the [apocalypse]" (2016). The short play set in a suburban garden sees the women talking about ordinary life events, which are interspersed with phobias, murder and abstract scenes—set beyond the garden—that depict, through monologues by Mrs J, pestilence, cannibalism, famine, death, despair, flood and the loss of culture and ideas, indeed perhaps even the capacity to think, through submission to constant social media products. For example, towards the end of the play, she describes the impact of fire on an unnamed society and ends with: "Finally the wind drove the fire to the ocean, where [saltwater] made survivors faint. The blackened area was declared a separate country with zero population, zero growth and zero politics. Charred stumps were salvaged for art and biscuits" (Churchill, 2016).

In other works, the focus is on imagery, the slow loss of connection and of meaning, and the ultimate destruction of the biosphere through humanity's craving for material resources. In Kris Verdonck's *Something (out of nothing)*, the work begins in darkness with a voiceover by Tawny Andersen, who states:

> It is impossible for me to move, awaken the dead, and make howl what has been broken. I can only see how the mountains of debris are piling up beneath me.
>
> Stuck. Stuck between past and future.
>
> Damage done.
>
> I, I am afraid of the rain, I am afraid of the air, I am afraid of the soil.
>
> The land.
>
> Look with me to the land, everything looks so familiar: the rocks, the trees, the animals, the air and the water. The beauty of it all only makes it worse. (Eckersall & van Baarle, 2020)

This voiceover is described by *Something (out of nothing)*'s dramaturge, Kristof van Baarle, as starting out in the work as "a more distanced narration of a human-made post-apocalyptic condition in the first part, moving to a witness-like perspective recounting the destructions that the voice saw in the second part, to end in a subjective more poetic speech position [in] the third part, highlighting the voice's lived experience of watching the catastrophes unfold" (Eckersall & van Baarle, 2020). This voiceover frames the work and is accompanied by scenes with fully masked actors in morph suits and business attire moving in fragmented and singular—or even atomised—ways, while several giant inflatable flower-like structures emerge and inflate from the ceiling. As van Baarle continues, "The difficulty was finding the right tone and perspective for what we wanted: as little human presence as possible, a focus on post-catastrophic moments, as straightforward and concrete as possible, the sincerity of the perspective of the witness and the objectivity of the journalist" (ibid.). They were influenced by myriad texts—literary, journalistic and philosophical—and, as is always the case with the work of Verdonck's A Two Dogs Company, Samuel Beckett. These different examples demonstrate, to some extent at least, the desire to deal with apocalyptic or tragic themes, but to render them in a way that eschews spectacle in favour of complex narrative and scenography that seeks to both obfuscate and uncover often painful, difficult and crucial ideas (Fig. 14.1).

Fig. 14.1 Photograph of *Something (out of nothing)*. (Image by Bas de Brouwer)

As a storytelling mode, film has the capacity to re-present extreme events that have already occurred and to imagine, in sound and image, future catastrophic scenarios. This capacity extends across fiction and nonfiction modes. Science fiction has been the primary fictional mode for rendering potent narratives about future dystopias. In the realm of nonfiction, landmark examples exist that offer a potent synthesis of the real and the speculative. Films such as Peter Watkins' *The War Game* (1965), Michael Madsen's *Into Eternity* (2010) and Juan Salazar's *Nightfall on Gaia* (2015) utilise documentary's address to bring extreme scenarios into the world of the viewer. The most notorious of these is the 1965 Oscar-winning feature documentary *The War Game*. Part of this infamy is due to the decision not to release the film as scheduled. Funded and produced by the BBC, it was deemed too horrifying for television broadcast. The film offers a representation of a nuclear attack occurring in British cities and employs the conventions of a television documentary to convey a sense of the attack occurring in real time. The film counts forward from the present, constructing a timeline into a fictional near future to reveal an unfolding narrative, including the physical effects on civilians. The dramatisation (or docu-drama) in *The War Game* offers an example of the way narrative conventions can be deployed in ways that potently locate the viewer in the world of the events taking place on screen. It is an instructive example and, albeit utilising analogue technology, one that anticipates the development of AI-supported images in contemporary filmmaking. As we have already noted, this move in film, gaming and other media allows for the visualisation of imagined scenarios in existing present-day contexts while also bringing the added potential of interactivity and multiple pathways or storylines, depending on the user.

Orthodox definitions understand film narrative as "a type of filmic organisation in which the parts relate to each other through a series of causally related events taking place in time [and space]" (Bordwell & Thompson, 2004). While historically film theory has positioned spectacle and narrative in contrasting ways, as storytelling platforms, the impact of a film is tied to the possibility for sensory engagement with the spectacle on screen (and its audio accompaniment). Adrian Ivakhiv summarises this relation by referring to spectacle as "cinema's firstness, [...] the immediate felt quality of the film image, the objectness, presentness, and thingness of what we see, hear and feel as we watch a film" (2013). Narrative, film's secondness, tends towards linearity and "provides us with the possibility of piecing together storylines as we make interpretive connections between the things we see and hear" (Ivakhiv, 2013). This orientation to narrative, if deployed skilfully, evokes a desire to experience what comes next. Together, spectacle and narrative offer a sensory experience, anchored in time and place, that structures worlds that we, as viewers, are invited to step into, and the promise of resolution, in part, maintains our connection to this world. In the case of dramatic re-construction or "pre-construction" (for an anticipated future scenario) of extreme events, film has the potential to prepare the viewer for what is to come by conveying the complexity, phenomenology and multiplicity of this experience.

14.3 Stories of Place and Community in Crisis: *Fires*

Fires (Ayres et al., 2021) is a six-part television series produced by the Australian Broadcasting Corporation and co-created by Tony Ayres and Belinda Chayko. The series was made in the wake of the 2019–2020 Black Summer fire season in eastern Australia. Each episode is loosely linked by the repeat appearances of two young volunteer firefighters, Mott (Hunter Page-Lochard) and Tash (Eliza Scanlen). While the characters are fictional and the places depicted largely unnamed, the intensity and characteristics of the firestorms and the landscapes in which they occurred accurately reflect the conditions and impacts of this fire season. For this reason, we position the series as a creative practice case study that might provide further knowledge about aesthetic and narrative strategies for preparing for extreme events. We explore these strategies from our different disciplinary perspectives, weighing up their function in terms of the series' illustrative and pedagogical power. Rather than following the original sequence of episodes, in this section, we start with the place of architecture in the landscape before focusing first on the experience of individuals during the events and finally on the traumatic interior landscape of emotions.

A series of slow, aerial pans across recently destroyed houses and blackened landscapes in New South Wales provides the opening scene for Episode four of *Fires*. It is Christmas Eve 2019, and residents of the isolated town of Bungan are being evacuated to a community hall as the fires get closer. This building provides the overarching narrative with a sense of place while also defining the bounds of a special type of community, temporary and strained in the face of a crisis. Bungan Hall is a place where people are thrown together, some meeting for the first time and others rekindling a past connection. It provides the mise-en-scène for much of the narrative, with the landscape as its ever-present, ominous counterpoint.

Bungan Hall is a weatherboard building made up of small, connected structures around a covered veranda. These are recognisable elements of Australian vernacular design, and, while faded and worn, the hall has clarity about it. In contrast, the landscape is typically indistinct and out-of-focus, a greenish-brown background to the hall. On the hall's rooftop, a man is shown staring into the distance, looking at the smoke from the closing fire front. It is obvious that he is clearing leaves from the gutters, to prevent embers from catching on the roof, allowing any firestorm to pass over. Volunteer firefighters also douse the exterior of the hall, in another well-understood attempt to stop the approaching fire from taking hold. Inside the hall, wet towels are placed at the base of doors to stop smoke from entering.

Overnight the fire passes by, and the episode concludes with a view of the community hall from directly above. The people who sheltered safely within are leaving to see if their homes are still standing. But as the view from above widens, the hall is revealed to be surrounded by a burnt landscape. This hall, with its distinct sense of place and Christmas trimmings, occupies a normative role in the narrative. It is a refuge of last resort, with a well-understood capacity to be defended from bushfires. There is, however, a clear counterpoint in the series, when architecture's failure to fulfil the traditional role of shelter is central to the narrative.

It is seven days later, New Year's Eve at Cooyang Point, a waterfront community in Victoria. This holiday town has been completely isolated by wildfires, and there are no avenues for escape. The local radio station advises people to evacuate their homes and make their way to the water. The town itself is rarely seen. There is a sense, amid the chaos, of a picturesque village with a main street lined with restaurants and shops. But the town and its architecture provide no refuge, and as the sun sets, the yellow-brown smoke-filled sky turns red and the surroundings burn. In this narrative, architecture is largely absent, unable to provide shelter or define a sense of place or community. A brief vignette dramatises this as a man tries to defend his home from the fire, but it is too late. The narrative then shifts to several months later, and people are shown living in cars or tents, and, tragically, to the north, towns are now being destroyed by floods.

These scenes resonate with an Australian audience, not only because Black Summer looms large in the public imagination but also because it came on the heels of an increasing number of megafires over the last two decades (Lindenmayer & Taylor, 2020, 12481). Through sound, image and narrative, *Fires* organises a range of human responses under various conditions that allow viewers to place themselves in the situations depicted with all the affective and experiential ramifications they entail. The combination of the actual historical setting and the realism of the genre of television drama makes *Fires* available for a screen studies analysis that draws on understandings of fiction and nonfiction forms.

Episode three is set in a coastal New South Wales town and follows two plotlines that show how individuals respond, and are situated or equipped, as fire advances and residents are advised to evacuate. Lally (Anna Torv) stays to continue preparations to defend the house as her partner Adrian travels to Canberra to drop his daughter off, planning to return before the fire hits. In another thread, recovering drug addict and campground employee Joel (Mark Leonard Winter) is left with no access to transport and without the methadone he relies on when the call to evacuate comes. He eventually persuades a family of campers to drive him out of the fire zone.

Lally's story attends to the subjective experience of being isolated and underprepared with a fire approaching. With Adrian gone, we follow her anxious movements in and around the house as signs of fire, such as burnt leaves blowing in on the wind, increase and day turns to night. Her initial confidence in defending the house (with a list of actions to tick off which the couple repeatedly refer to) soon wanes as the house becomes a place of uncertainty. The smoke haze increases and flying foxes, disoriented by smoke, are in a state of high activity, with one crashing into a windowpane. The only resident in the close vicinity who chose to stay, Lally is isolated. We are asked to identify with her heightened state of anxious waiting. There are indications that she is questioning their plan. She glances at her packed suitcases and makes intermittent calls to Adrian, imploring him to return soon. He becomes delayed at a roadblock and Lally seals one room off from the smoke with towels after breaking into a neighbour's house to retrieve batteries. Adrian returns in the light of day, driving up to a house that has, so far, survived the fire. In a tense conversation, Lally asserts that she will leave while she can, no longer willing to risk the uncertainty of the situation.

This plotline is affective rather than simply instructional. It does not explicitly illustrate best practice for fire preparedness, but it does present a subjective and sensory scenario that highlights the complexity and nuances of fire preparedness for communities and individuals. The focus on characters distinguished by gender and/or social status draws attention to the diverse conditions for decision making during an evacuation. These distinct social subjectivities also serve to highlight the importance of relations between individuals. Lally and Adrian are separated and their plans become derailed. The final shot of Adrian as Lally drives away shows him standing on the roof of the house, hosing it down, determined to stay. This contrasts starkly with Lally's vulnerability the previous night. Working at the campground, Jeff, alone and underresourced, had no plan and was left to rely on the generosity of others. Both groups of characters are diverted from their plans, with road closures and their locations in areas relative to the fire important factors in the creation of a causal, temporal and spatial coherence across the world of the episode.

With time and space as constituent factors, people are separated or thrown together in unpredictable ways, showing that social relationships are crucial and yet can be easily disrupted in extreme events. Because episodes in *Fires* typically unfold over the course of a single day, they allow us to observe the experience of moving from confidence and security to panic. They offer the opportunity for the viewer to step into a nonfiction world that has a strong relationship with what the viewer knows about the past and possible future experiences of actual individuals. They show how the moving image in narrative structure can shape thinking about the temporality of fire and firestorms, especially as an unpredictable and sensory experience. They also prepare people to think about the aftermath of the event, the wreckage of lives, animals, livelihoods and landscape.

Episode two begins with the Simpsons, Kath (Miranda Otto) and Duncan (Richard Roxburgh), returning from evacuation after the fire destroyed their homestead, to have dinner and the opportunity to rest with their neighbours. Kath and Duncan are accompanied by their missing son's partner Brooke (Taylor Ferguson)—who remained at the house she shares with her partner Lochie, their eldest son, on the family farm. The tone is sombre and the sense of stoicism that the Australian farming and grazing community tends to be characterised by is clearly apparent.

This episode centres on the gradual realisation that Lochie, who they have not been able to contact since the fire, and who did not return once the road was opened, might be among the dead. There is tension between Kath and Brooke that permeates all their early scenes. It is characterised by a sense of cultural difference. Kath admits that she's "never taken Brooke and Lochie seriously" in terms of their relationship. She sees Brooke as a city girl, not really invested in dairy farming. Each character battles grief in their own way and Kath's apparent inability to show vulnerability in the face of trauma ensures that the unfolding story is a profoundly sad experience. The day-to-day farming jobs that must continue despite the enormity of the losses—milking the dairy cows, cleaning out stalls, etc.—are performed in a silent and robotic fashion by the distraught Duncan and Kath.

Fig. 14.2 Still image from Ep. 2 Fires by ABC TV/Matchbox Pictures. All rights reserved

The entire episode is emotionally taut and only allows a few small moments of release for the characters and the viewers. The moment Duncan finds Lochie's burnt-out ute in the back roads is deeply moving. We follow him as he tries to retrace Lochie's movements, mapping out road closures, thinking through how Lochie might have responded to approaching fires and wind, saying out loud to himself "where did you go", and then he finds the truck, a white carcass in the midst of smoky, black, burnt-out trees and ashen landscape. Duncan realises that Lochie was trying to take a shortcut to get back to the farm with a pump to assist with the efforts to fight the fires and protect the farm and Brooke, while not realising that his parents had been evacuated. The scale and ferocity of the fire is rendered clearly in both the charred landscape and the figure of Duncan as he negotiates the scene.

While this episode differs in its focus on the aftermath of the fires rather than on the acts of preparing, evacuating and experiencing them, it plays an important part, as it gives a roundedness to the series in attempting to invite audiences in to see and understand that the impacts of devastation and loss are myriad and that they happen to complex, messy and ongoing lives—the lives of people with stories, with pain and with unresolved relationships.

14.4 Forging New Narratives for a New Future

One message that can be seen in the aftermath of years of ever-increasing fires and floods is that our narratives and practices around responding to climate emergencies must evolve. To consider this need, we provide some brief perspectives on how this might be realised from our disciplines and how applications of advanced media and technology may support this goal.

In terms of thinking about the evolution of artistic practices in theatre and performance, it is easier to see how artists can work on the ground with communities either pre- or post-disaster, than to think about the role of theatre and performance more broadly (beyond the applied or community context) in adapting to, or preparing for, extreme events. While there are examples of innovation, some of which we have discussed in this chapter, a focus on developing practice that draws on advances in technology to imaginatively envision or render scenarios and therefore assist spectators in gaining a deeper understanding of these is crucial. For those artists and companies interested in ecological art or performance in the context of a rapidly changing climate, the advantage of collaborating with digital artists is one that offers myriad possibilities for opening how we see, experience and understand this new "normal". As examples described in this chapter demonstrate, the immersive potentiality of these technologies allows for new kinds of spectatorial encounters—indeed, a move beyond spectatorship into a process of participation in the unfolding of experiences, events and scenarios. Engaging with these technologies will transform the potentiality of performance to deal with complex and difficult topics but will also expand the opportunity for deep understandings of what an extreme weather scenario is or can be.

As an expansion of this idea, the documentary, and especially the docudrama, provides opportunities to explore how "pre-construction" (as opposed to re-construction) might be considered a type of adaptive engagement. Such an approach would require close attention to ideas of narrative and spectacle and their joint capacity to engage the audience in a deep sense of place and time. Technology, not just in the production of the moving image but also in the way it is experienced, offers a way to heighten the influence of the documentary as a pre-constructive mode. Immersive simulations also offer a new way to experience documentary. Moreover, rethinking our settled assumptions about time and the moving image has significant potential to help retrain our perception and thus responses to extreme events. One avenue for exploration is nonlinear interaction and instances where people can reshape the sequence of the narrative, individualising both the experience and its emotional impact. Landmark interactive documentaries (or I-docs) which ask viewers to engage with more-than-human worlds include *Bear 71* (2012), *The Shore Line* (2017) and *The Whale Hunt* (2007). These offer an important starting point for conceptualising this role for documentary. Elizabeth Miller's pioneering works, *SwampScapes* (2018) and *As the Gull Flies* (2023), are examples that explore documentary's potential in VR and AR form. Yet another exploration might interrogate individuals' perception of time, adjusting the duration of events and cause and effect relationships to productively customise perceptual adaptation for individuals and groups.

Architecture plays a complex role in the imagining of these immersive possibilities for place and community. It is neither as fixed nor as anodyne as it is often portrayed on film, in books and in computer games. Furthermore, in the modern world, and in contemporary disaster preparedness practice, we can no longer treat buildings and landscapes as immutable; they are active and often unpredictable components of our responses to natural disasters. Future research is needed to envision what a less passive architecture or a nonlinear, interactive exploration of the landscape would bring to the creation or maintenance of a sense of place or community as sources of both protection and danger. Simulations using advanced technology, as either interactive digital narrative or cooperative computer games, can provide a laboratory for developing these ideas, while creative hubs in the community offer a way of implementing or disseminating the new knowledge they produce, supporting the evolution of our current models of practice into a new, more dynamic model for the future.

14.5 Conclusion

For Casey, "the climate crisis is the ultimate crisis: it threatens to destroy life in the oceans and the atmosphere as well as on land. It presents us with a circumstance in which biological life, always fragile and on edge, will be faced with a new and final challenge: how to remain robust enough to survive the [extremes] of climate chaos" (2017). In this dire situation, we return to the focus of this chapter, which is on how our different disciplines imaginatively engage not only with the status quo but with creating works that assist communities to build resilience, to plan for and consider in advance the extreme scenarios that are likely to happen. To do this in the context of theatre studies, screen studies and architecture is, we believe, to engage in a process that supports communities to narrate their present and imagined futures in ways that empower them to be "robust enough to survive". This narration needs to be multifaceted. It must accommodate localised interventions with vulnerable groups as well as avant-garde imaginings of scenarios and events. It must also adapt and be relevant to contemporary audiences by drawing on key technological advances in VR and AI, in developing works that are immersive, stimulating and engaging and that can involve them in imagining future scenarios and, therefore, position them to think about how they must act to survive in places vulnerable to extreme events.

References

Ayres, T., Chayko, B., Denholm, A., & Watts, L. (executive producers). (2021). *Fires* [TV Series]. Australian Broadcasting Corporation.
Bordwell, D., & Thompson, K. (2004). *Film art: An introduction*. McGraw-Hill.
Casey, E. S. (1998). *The fate of place: A philosophical history*. UCP.
Casey, E. S. (2001). Between geography and philosophy: What does it mean to be in the place-world? *Annals of the Association of American Geographers, 91*(4), 683–693.
Casey, E. S. (2009). *Getting back into place: Toward a renewed understanding of the place-world* (2nd ed.). Indiana University Press.
Casey, E. S. (2017). *The world on edge*. IUP.
Churchill, C. (2016). *Escaped alone*. Nick Hern Books.
Clapp, S. (2016, January 31). *Escaped alone* review – Small talk and everyday terror from Caryl Churchill. *The Guardian*. https://t1p.de/6h0o2. Accessed 17 Oct 2023.
Cowie, E. (2011). Documentary space, place, and landscape. *Media Fields Journal, 3*, 1–21.
Eckersall, P., & van Baarle, K. (2020). *Machine made silence: The art of Kris Verdonck*. Performance Research Books.
Ivakhiv, A. (2013). *Ecologies of the moving image*. Wilfrid Laurier UP.
Jacobs, J. (1961). *The death and life of great American cities*. Random House.
Lindenmayer, D., & Taylor, C. (2020). New spatial analyses of Australian wildfires highlight the need for new fire, resource, and conservation policies. *Proceedings of the National Academy of Sciences, 117*(22), 12481–12485.
Miller, E., Little, E., & High, S. (2017). *Going public*. UBCP.
Norberg-Schulz, C. (1980). *Towards a phenomenology of architecture*. Rizzoli.
Salazar, J. F. (2015). Anticipatory modes of futuring planetary change in documentary film. In A. Juhasz & A. Lebow (Eds.), *A companion to contemporary documentary film* (pp. 43–60). Wiley Blackwell.
Smaill, B. (2016). *Animals and the documentary moving image*. SUNY Press.
Walker, J. (2005). *Trauma cinema: Documenting incest and the Holocaust*. UCP.
Whyte, W. (1980). *The social life of small urban spaces*. Conserv. Foundation.

Chapter 15
Communicating in Crisis: Community Practices of Online Participation During Extreme Events

Sora Park, Susan Atkinson, Janet Fulton, Gabrielle Wong-Parodi, and Lara Mani

Abstract This chapter surveys research into the communication among community members affected by extreme events with digital platforms such as social media and messaging apps before, during and after the events. While there is extant literature on how people adopt effective strategies in sharing real-time information during a major crisis, fewer studies examine the entirety of the process, particularly around preparing communities and individuals, and even fewer focus on how community members seek and share social support. This chapter examines both aspects of digital communication—emotive and informative—to better understand the role digital platforms can play in extreme events in supporting more effective responses. It also identifies gaps in the literature on the role of social media in preparing individuals and communities for catastrophic climate events.

Keywords Communication · Community participation · Crisis communication · Digital participation · Disaster preparedness · Natural hazards · Resilience · Social media

---

S. Park (✉) · S. Atkinson
News and Media Research Centre, University of Canberra (UoC), Bruce, ACT, Australia
e-mail: sora.park@canberra.edu.au; Sue.Atkinson@canberra.edu.au

J. Fulton
College of Design & Social Context, RMIT University, Melbourne, VIC, Australia
e-mail: janet.fulton@canberra.edu.au

G. Wong-Parodi
Stanford Woods Institute for the Environment, Stanford University, Stanford, CA, USA
e-mail: gwongpar@stanford.edu

L. Mani
Centre for the Study of Existential Risk, Cambridge University, Cambridge, UK
e-mail: lm881@cam.ac.uk

D. Del Favero et al. (eds.), *Climate Disaster Preparedness*, Arts, Research, Innovation and Society, https://doi.org/10.1007/978-3-031-56114-6_15

15.1 Introduction

Increasing emergencies and disasters in Australia and worldwide are leading to the realisation that response agencies do not have the resources to be everywhere and that communities have been (and always will be) the first responders in any emergency or crisis. Recent Australian and international events (earthquakes, wildfires and floods) have highlighted wide discrepancies in the capabilities of communities to adequately respond before the arrival (and, in many cases, non-arrival) of emergency and rescue agencies. Australia's size and population spread exacerbate this situation and mean that local community communication networks play an essential role in emergency responses as well as preparedness. The complex emergency media ecosystem in Australia sees an interweaving of different media channels, including the national public broadcaster the Australian Broadcasting Corporation (ABC), commercial local media and direct communication from the emergency sector websites and apps. Communication occurs via traditional media channels, such as radio as well as SMS and social media. Due to its immediacy, social media platforms play a crucial role "in the dissemination of information and the coordination of community responses" (Bruns, 2014).

The role that community members play in emergencies has become increasingly visible through the use of digital technologies and social media. A recent comprehensive report into how people experienced Australia's damaging floods in early 2022 clearly displayed the changing role of communities in responding to and managing emergencies. It emphasised a move away from reliance on official channels towards established and trusted community sources on social media to meet the need for timely, accurate and relevant local information and to provide and receive support and assistance:

> Many people [rely] on local knowledge—[whether] their own, or that of others—to contextualise the potential impacts they might face and to take protective action…There are increasingly well-informed, experienced, and knowledgeable individuals and groups working through informal networks and basing their information on evidence, technology, and modelling. (Taylor et al., 2023)

One participant in the Taylor et al. report stated: "I learnt more from our [community] Facebook group than I did from watching the news and trying to get weather reports and stuff" (2023).

This chapter examines how people use these technologies and platforms not just for information-seeking but to fulfil other needs, such as to create and share meaningful, relevant hyperlocal information, seek and offer social and emotional support and coordinate immediate and local response activities (Austin et al., 2012; Bruns, 2014). It will also attempt to address the question of preparedness in the context of crisis communication and identify knowledge gaps and future opportunities for research: What do we not know and what can we do? What has been discovered in the research literature is a heavily skewed focus on response and recovery, with very little on what can be done from a crisis communication perspective before an event occurs to prepare communities and individuals. Ng's (2022) definition of disaster preparedness is helpful here, drawing on Paton (2019) and Dasgupta et al. (2020), to "refer[–] to activities and measures taken in advance to ensure an effective response to the impact of hazards". Social media has become an integral part of our daily lives. Globally 95% of adults use social media on a weekly basis, and in Australia, this figure is 92% (Newman et al., 2023; Park et al., 2023). In this chapter, we refer to all types of digital platforms that enable interactions between individuals to communicate and share information as social media, mainly focusing on social networking sites (Meta, X, Instagram, etc.) and instant messaging applications (WhatsApp, Facebook Messenger, WeChat, etc.). Due to the pervasiveness of social media platforms, it has the potential to disseminate essential official information during natural disasters as well as serve as a platform to exchange information and support among citizens. In contemporary emergency management practice, the term "natural hazards" is preferred over "natural disasters" as all hazards (including security-related events, public demonstrations or major IT incidents, as well as floods, wildfires and heatwaves) are managed within the same framework. In the crisis communication field, the term "natural disasters" is used when discussing crisis events that have an environmental genesis. As we straddle both these fields of research, the terms are used interchangeably in this chapter.

During an emergency, people encounter a large volume of information, including unverified information. Filtering accurate and relevant information is challenging as social media facilitates inadvertent as well as deliberate spread of inaccurate information—although this has been found to be less likely during a disaster (Palen et al., 2009; Simon et al., 2015). Some research has shown that social media often serves as a self-regulating device where users fact-check and share the corrected information (Simon et al., 2015).

While access to timely and accurate information is a critical component of emergency response, and successful communication undoubtedly contributes to better outcomes (Steelman & McCaffrey, 2012), it is only part of the picture. For individuals, sourcing and sharing information to make better-informed decisions is essential. Furthermore, during emergencies, people engage with and communicate with others for many other reasons, the importance of which can be lost if focus is narrowly concentrated on content, messages, channels and information flows. These reasons include understanding local conditions by talking with others; making judgements about what specific information might mean for them; coordinating assistance for themselves and others; offering and receiving emotional support; and making people feel safer through personal connection (Sharp & Carter, 2020; Simon et al., 2015; Stephens & Robertson, 2022). Engagement and communication, particularly through social media, are key elements in building community cohesion and connectedness, which research has determined to be essential for preparedness, response and recovery from disasters and in building what has been defined as community resilience (Liu, 2022).

The use of social media and "unofficial" media is not a new phenomenon, but, increasingly, there is an expectation of immediate, local information and a move towards established and trusted community sources (Taylor et al., 2023). People value local knowledge and information as it enables them to contextualise the potential impacts of a crisis, make their own decisions and take necessary actions. Hyperlocal knowledge is often more relevant to individuals as well. When official information is absent, perceived as inaccurate or lacking in some way, people rely on unofficial sources of information such as community Facebook groups and their own personal networks for real-time localised information (Stephens & Robertson, 2022; Taylor et al., 2023; Zander et al., 2022). These communities can also be maintained during non-emergency periods for citizens to maintain connections and prepare for the next disaster.

This chapter includes a review of scholarly research on the role of social media in emergency communication to identify the insights and gaps in understanding how social media might be used in preparing individuals and communities for natural disasters.

15.2 Social Media and the Changing Communication Landscape During Disasters

15.2.1 Social Media as an Expansion of Official Communication Channels Assisting Emergency Communication

Research consistently shows that citizens "perceive official sources, such as [government] agencies, as more credible for disaster information than unofficial sources, such as members of the public, both via traditional and social media" (Fisher Liu et al., 2016). People tend to "use social media during crises for insider information and to check in with family and friends, while they use traditional media for educational purposes, i.e., seeking official information and warnings" (Sheldon, 2018). However, people "typically use a combination of unofficial and official sources to make sense of disaster information" (Fisher Liu et al., 2016). Here, social media can play an important role in seeking trusted sources of information. During disasters, people seek information to locate family and friends and reduce uncertainty. For this, they usually turn to familiar channels like phone calls, text messages and emails to gather information, and, if unsuccessful, they resort to alternative and official sources of information (Simon et al., 2015).

Research by Bruns and others (notably Austin et al., 2012; Fisher Liu et al., 2016; Liu, 2022) has emphasised that the unique characteristic of social media, its capacity to facilitate immediate transmission from many-to-many, is why it must be included in emergency agencies' communication strategies, but not just as one-way communication channels. However, research has shown that responders (i.e. emergency authorities) prefer to receive information rather than share it, making it challenging to coordinate and share information and missing opportunities to engage citizens in information gathering and sharing (Simon et al., 2015).

From a practitioner perspective, many studies focus on effective uses of social media among crisis communication professionals (see Eriksson, 2018; Eriksson &

Olsson, 2016; Maal & Wilson-North, 2019). Eriksson (2018), for example, recommends five effective uses of social media in crisis communication: (1) "[create dialogue and] choose the right message, source and timing"; (2) pre-crisis work; (3) "social media monitoring"; (4) "continuing to prioritize traditional media in crisis situations"; and (5) "strategic crisis communication". Note that Eriksson's model includes pre-crisis work, a step that is crucial in preparing for an extreme event, although Eriksson is discussing how communication professionals prepare rather than a community or individual.

Bruns (2014) emphasises the critical role of social media in further disseminating "emergency alerts and other messages and in the community self-organisation of local responses". As well as facilitating the dissemination of information from traditional emergency management actors, social media facilitates the emergence of new actors such as non-governmental organisations (NGOs) and other civic organisations, active Internet actors and locals sharing first-hand information or organising local emergency response activities.

However, the affordances of social media platforms come with challenges. Maal and Wilson-North (2019) note the amplifier effect that social media has on crisis communication and the challenges in ensuring accurate messages are heard and unverified information is confirmed or refuted. Their guidelines for using social media in crisis communication encourage organisations to carefully consider how people use different platforms and tailor communication strategies accordingly. In the Western context, for example, X (formerly Twitter) may be used for daily safety messages, incident information and retweets of key messages from partner organisations, while Facebook should be used for more detailed stories and safety campaigns. Maal and Wilson-North (2019) recommend testing message formats on various channels and monitoring engagement statistics to evaluate different approaches.

There is very little control over messages once they are distributed on social media. While organisations can target audiences with their own messages, they do not know what other sources people are using and what other information people are seeking (Stephens & Robertson, 2022). In discussing the different affordances of social media platforms, they differentiate between public, semi-public and shared-private profiles that people use to connect with others. A significant challenge that official emergency communicators face is an increase in the public's belief that the official organisations are monitoring their social media and will respond. In reality, response organisations often lack the resources and training to effectively monitor these platforms, leading to a disconnect between public expectations and reality (Stephens & Robertson, 2022).

There is increasing demand for localised and real-time information before and during disasters, and research is finding a shift towards relying on community-curated information on social media platforms (Taylor et al., 2023). However, there are more studies that focus on how official agencies and practitioners use social media as a communication channel than on how people actually use social media to find, co-create and share information to support each other during crises (Macnamara, 2016).

15.2.2 Social Media Use by the Public During Disasters

For individuals and communities, social media is much more than an information source. Research has shown that in addition to searching for official information such as weather updates, river gauge heights and fire ground information, individuals use local online communities to find and share hyperlocal information. They do so to maintain social connection with friends, family and community members and to coordinate response and support activities—all activities that contribute to greater community cohesion and resilience (Liu, 2022; Sharp & Carter, 2020; Taylor et al., 2023). As Simon et al. (2015) claim, as well as becoming an essential conduit for information gathering and sharing, social media provides a platform for public participation and civilian journalism, enabling individuals and groups to seek, share and synthesise information.

Research going back a decade highlights the ongoing need to understand better how citizens use different social media platforms in times of crisis (Fraustino et al., 2012). This has been echoed by many scholars since (Eriksson & Olsson, 2016; Liu, 2022; Simon et al., 2015). As well as a communication channel, "social media provides opportunities for engaging citizens in emergency management by disseminating information [to a broader network] and accessing information from them" (Edwards et al., 2022). The importance of community connectedness and the diverse roles played by friends, family and neighbours during disasters have been highlighted by various studies (Liu, 2022; Simon et al., 2015; Spialek & Houston, 2019; Taylor et al., 2023). These community members are the first responders, alerting others to the threat of flooding, sharing vital information and providing support in numerous ways such as assisting with evacuations, participating in clean-up efforts and aiding in recovery. Taylor et al.'s study revealed that community members "were identified as the most trusted and the most appreciated groups" and that "community cohesion and connectedness acts as a much-needed force multiplier when it comes to disasters" (2023).

Social media use has a clear value to communities during emergency events. For example, crowdsourcing information during an emergency enables "human sensors" who can regularly and rapidly update the situation on the ground, which may not be possible via official channels (Bruns, 2014). Due to these affordances, research finds a shift towards relying on community-curated information on social media platforms (Taylor et al., 2023). Atkinson and Lee's (2023) study on how the general public utilised the official pages of emergency organisations during natural disasters reveals that users engage with different types of content in different ways and that community-related information, as opposed to event-specific information, generated the highest levels of engagement. Taylor et al. (2012) found that during tropical cyclone Yasi, the public relied on a mix of formal and informal information sources, often using social media to re-post information from government websites that they felt would be useful to other people. Individuals were acting as "filters and amplifiers of official information". The core strength of social media use during a crisis is its "timely information exchange and promotion of connectedness". The study found that social media was used as "psychological first aid in the early stages of disaster", providing support among the community.

Community-curated content on social media, particularly Facebook, played a significant role in the catastrophic floods in Northern New South Wales and Queensland in 2022, with several Facebook pages in flood-affected regions emerging as reliable sources of high-quality information upon which residents increasingly relied. These pages effectively shared and amplified official emergency messaging, highlighting the potential for better collaboration between official agencies and community groups (Taylor et al., 2023).

Taylor et al.'s (2023) flood report stated that almost two thirds of respondents to the survey indicated that they used social media, with the most preferred sites being local community groups followed by official sources such as Queensland Fire and Emergency Services (QFES), New South Wales State Emergency Services (NSW SES) and the Bureau of Meteorology (BOM). Taylor et al.'s report quotes residents who found community Facebook pages to be centralised points of information where major updates were pinned, and individuals with direct access to the affected areas could provide real-time updates. Critical to these pages working well were the page administrators who did their best to validate information and promote the most recent information, including from official sources. Studies have shown that people often prefer this type of content because of its timeliness, accuracy, local relevance, community connection, use of visuals/photos and availability.

Taylor et al.'s report highlights a loss of trust in institutions and a growing "[preference] for relying on community-generated information on social media and local knowledge". This shift in trust and reliance underscores the need for effective engagement with community social media platforms and the incorporation of local knowledge into formal emergency management systems.

The content that ordinary citizens create and share during a crisis, however, is not solely about information sharing. People create content that "is tied to social [functions] and ego-defensive functions, [and] not to utilitarian knowledge functions" (Austin et al., 2012). Austin et al. found that "young adults' motivations for use of newer communication technologies is [most] related to their need for connectedness", for "self-expression" and, to a lesser extent, for "utilitarian purposes" (ibid.). People's differing motivations for using social media have implications for channel selection, type of content sought and perceived usefulness of a specific platform or group, and these may change radically during a crisis to meet changing needs. What has also been found is that in some cases, information overload may discourage the use of social media when people feel overwhelmed by the frequency and amount of information in their feeds (Austin et al., 2012). Generally, though, audiences' social media use increases during crises, and, "in some situations, audiences perceive social media to be more credible than [traditional] media", particularly for hyperlocal information (Austin et al., 2012). Using social media beyond information sharing, such as gaining social support, has the potential to enable stronger ties within local communities.

15.3 The Role of Social Media Before, During and After Disasters

While there is research around the role of social media during and after disasters, there is less about how it can be used, or is used, before an event. If preparedness is noted, it is typically brief. Nevertheless, in Zander et al.'s (2022) examination of how Australians use social media during natural disasters, they also asked their respondents about preparedness. While the research primarily covered individual and community use of social media *during* a crisis, they briefly noted household preparedness. Zander et al. claim that when information is given to households on how to prepare, this information can motivate individuals to prepare more effectively and recover more quickly. In this research, it was found that "[Being] acquainted with local emergency systems was positively associated with the use of social media during natural hazards" (2022), with the opposite also true: If a citizen depended on information from authorities, it was more likely to have a negative effect on their use of social media. In other words, those who trust the authorities do not feel the need to search out information elsewhere, and those who doubt the decision makers search for other communication sources: in this case, according to Zander et al. (2022), social media. It should also be noted that Zander et al.'s research included a literature review of research that examined how Australians use social media during emergencies (21 studies in total), and none of these investigated how social media was used as a tool for preparedness. Sharp and Carter (2020), in research that examined "the role of social media in preparation for, and response to, flooding", also primarily focused on response but made the observation that social media can strengthen community cohesion, implying that a strong community is more resilient and, thus, able to better manage risk during a disaster, an observation consistent with other researchers such as Liu (2022), Sharp and Carter (2020) and Taylor et al. (2023). Other research (e.g. Allaire, 2016) use *prepare* in the context of what a household can do during an event: "With knowledge of current flood conditions, social media households could prepare effectively and successfully move their [belongings] in time".

In 2014, Houston et al. developed a framework to map disaster social media by conducting a review of literature. This review generated 15 social media uses, including two pre-event phases: preparedness information, both developing and receiving it, and disaster warnings. Houston et al. note that social media can connect organisations and governments to individuals to disseminate information for households to prepare. They also point out that social media can lead to individuals unintentionally learning about preparedness from their feeds:

> [A] user may encounter disaster preparedness information via a posting from an account they follow on Facebook or Twitter (one that is not specifically disaster-focused, but has still posted or tweeted the information), thereby expanding the effect of disaster preparedness information beyond those who are motivated to look for the content. (Houston et al., 2014)

There is no doubt that social media has "changed the traditional information dissemination pathways during emergencies", resulting in "many more information providers and higher involvement of the public using official and unofficial sources" (Simon et al., 2015). This provides both challenges and opportunities (Bruns, 2014; Houston et al., 2014). Lui (2022) quotes Spialek et al. when noting that "social media are rapidly becoming an alternative for individuals to acquire and disseminate disaster-related information. In fact, [compared to traditional] media, social media can contribute to the building of community resilience to a greater degree". They find that "social media enable real-time updates of a disaster, making residents not only recipients but also [creators and] providers of information" (Liu, 2022). However, a key challenge is that disinformation and inaccuracy can spread more rapidly than via traditional media. All the while, emergency responders face challenges in validating information accessed by the public (Mehta et al., 2017; Simon et al., 2015; Stephens & Robertson, 2022).

At the community level, the disaster communication ecology is complex and multilayered, and official emergency management communication is only one component. Community-level "communication resources are distributed at multiple levels" within a community, and "the strength of communication ecology can be enhanced or restricted by various characteristics of [neighbourhood] context" (Liu, 2022). These "communication resources [include] interpersonal connections, local media storytelling and community-based organizations, and official emergency management communication" (ibid.). Different individuals rely on these resources differently, depending on their goals. Austin, Liu and Jin note that "[during] crises, social media allows users to share unfiltered, [immediate and local] information and news, which provide unique crisis information that audiences cannot get elsewhere. Audiences also use social media for emotional support and recovery from crises".

Liu (2022) proposes a four-component model for disaster assessment and intervention, which emphasises four key elements that are essential for effective disaster response. The first component is robust communication systems, which play a critical role in coordinating relief efforts and keeping the affected population informed. The second component is strong community relationships, which can facilitate the sharing of information and resources and promote social cohesion. The third component is positive community attributes, which are essential for building resilience and fostering a sense of community ownership over the relief and recovery process. Finally, the fourth component is effective strategic communication processes, which can help to coordinate the efforts of different stakeholders and ensure that the right information is conveyed to the right people at the right time.

Here, the role of interpersonal communication is a vital component of disaster communication ecology (Kim & Kang, 2010). Talking to family, friends and neighbours can serve several important functions. People tend to trust information from family and friends and it is often hyperlocal. Another important function of interpersonal communication is that it can lead to collective action in the community, which is critical during preparation for disasters and recovery from them. Finally, connecting with others during a disaster fosters a sense of connectedness and can encourage community activities (Liu, 2022).

Studies have shown that interpersonal disaster communication can reinforce the effects of news media and enhance community resilience perception, collective efficacy and disaster self-efficacy. Interpersonal communication combined with news media can increase individuals' engagement before and during disasters (Kim & Kang, 2010). Similarly, Spialek and Houston's (2019) study found that people's disaster communication activities were related to a sense of belonging and resilience, confirming the important role of interpersonal disaster communication in shaping community resilience.

15.4 Conclusion

A key inference from research into this area of crisis communication is "that social media [should] not [be considered] a homogenous phenomenon with [one] coherent role in crisis management and communication [practice]" (Atkinson & Lee, 2023). Communication professionals and citizens assign important distinguishing characteristics to different social media platforms. Groups diverging in profession, age, gender, interests and experience will likely continue assigning different roles to diverse social media platforms, both in everyday use and during crisis events. This variability will influence how community communication networks are understood and how authorities may develop their strategies.

Most of the literature examined here focuses on organisations and conceptualises communication as one-way message delivery, relegating social media to something that complements the existing media and communication systems. It is, however, prudent to consider social media from multiple perspectives. From an organisation's perspective, social media has demonstrated its usefulness in extending official channels of communication and has the potential to be a source of situational intelligence and a platform to engage with communities. From an individual's perspective, social media can be an effective tool to manage the flow of information and social support and coordinate response efforts. Credibility is higher for official social media sources (Fisher Liu et al., 2016) than for unknown sources, although it must be argued that trust is high for interpersonal information, mainly local information.

However, a key finding is how important it is to build community preparedness, capability and resilience and how significant intra-community communication is as an enabler. Much of the literature examines strategies to share information during a crisis, with very little done on how individuals and communities can effectively use platforms such as social media, digital technologies and messaging apps to manage communication before, during and after an event. Here, we have uncovered several gaps in the research, particularly on preparedness, but also around concepts such as the value of different information sources, and how social media can be used after an event to prepare for future emergencies and for emotional support during disasters. In addition, digital technologies will continue to advance and provide more opportunities to enhance community connectedness and preparedness, potentially including through the creation of community hubs where diverse communication networks could come together. These gaps should be explored empirically, including via longitudinal studies, to provide further data on improving preparedness for individuals and communities.

The following gaps were identified:

- People prefer interpersonal communication and are less willing to share information online (Fisher Liu et al., 2016). They will use a combination of official and unofficial information sources. There needs to be more visibility of the unofficial and spontaneous information sharing. This preference can make it challenging to research but is worth exploring empirically.
- Preparedness is not a core element of these studies; most studies are about how social media is used during response and recovery. A closer examination of how social media can be used to assist in preparing communities and individuals for catastrophic events would add a further dimension to emergency management research.
- Existing research points to variations in how digital technologies are used in different crises, such as floods, wildfires and heatwaves; however, empirical research is required to offer a more nuanced understanding of these differences.
- It is largely unknown how social media communities and networks established during disasters continue to evolve post-crisis. Empirical research can question and examine the ongoing value of these communities.
- There are very few studies on emotional support, which is a part of resilience building and preparedness.

References

Allaire, M. C. (2016). Disaster loss and social media: Can online information increase flood resilience? *Water Resources Research, 52*(9), 7408–7423.

Atkinson, S., & Young Lee, J. (2023). Social media: Connecting and sharing in a bushfire crisis. *Media International Australia*. https://doi.org/10.1177/1329878X231163367

Austin, L., Fisher Liu, B., & Jin, Y. (2012). How audiences seek out crisis information: Exploring the social-mediated crisis communication model. *Journal of Applied Communication Research, 40*(2), 188–207.

Bruns, A. (2014). Crisis communication. In S. Cunningham & S. Turnbull (Eds.), *The media and communications in Australia* (pp. 351–355). Allen & Unwin.

Dasgupta, R., Basu, M., Kumar, P., Johnson, B. A., et al. (2020). A rapid indicator-based assessment of foreign resident preparedness in Japan during Typhoon Hagibis. *International Journal of Disaster Risk Reduction, 51*, 1018491.

Edwards, F., Liu, K., Lee Hughes, A., Gao, J. Z., Goodrich, D., Barner, A., & Herrera, R. (2022). *Best practices in disaster public communications*. Mineta Transport. Institute.

Eriksson, M. (2018). Lessons for crisis communication on social media: A systematic review of what research tells the practice. *International Journal of Strategic Communication, 12*(5), 526–551.

Eriksson, M., & Olsson, E.-K. (2016). Facebook and Twitter in crisis communication: A comparative study of crisis communication professionals and citizens. *Journal of Contingencies & Crisis Management, 24*(4), 198–209.

Fisher Liu, B., Fraustino, J. D., & Jin, Y. (2016). Social media use during disasters: How information form and source influence intended behavioral responses. *Communication Research, 43*(5), 626–646.

Fraustino, J. D., Fisher Liu, B., & Jin, Y. (2012). *Social media use during disasters: A review of the knowledge base and gaps*. START.

Houston, J. B., Hawthorne, J., Perreault, M. F., Park, E. H., Goldstein Hode, M., Halliwell, M. R., Turner McGowen, S. E., Davis, R., Vaid, S., McElderry, J. A., & Griffith, S. A. (2014). Social media and disasters: A functional framework for social media use in disaster planning, response, and research. *Disasters, 39*(1), 1–22.

Kim, Y.-C., & Kang, J. (2010). Communication, neighbourhood belonging and household hurricane preparedness. *Disasters, 34*(2), 470–488.

Liu, W. (2022). Disaster communication ecology in multiethnic communities: Understanding disaster coping and community resilience from a communication resource approach. *Journal of International and Intercultural Communication, 15*(1), 94–117.

Maal, M., & Wilson-North, M. (2019). Social media in crisis communication—The 'do's' and 'don'ts'. *International Journal of Disaster Resilience in the Built Environment, 10*(5), 379–391.

Macnamara, J. (2016). The work and 'architecture of listening': Addressing gaps in organization–public communication. *International Journal of Strategic Communication, 10*(2), 133–148.

Mehta, A. M., Bruns, A., & Newton, J. (2017). Trust, but verify: Social media models for disaster management. *Disasters, 41*(3), 549–565.

Newman, N., Fletcher, R., Eddy, K., Robertson, C. T., & Kleis Nielsen, R. (2023). *Reuters Institute Digital News Report 2023*. UOP.

Ng, S. L. (2022). Effects of risk perception on disaster preparedness toward typhoons: An application of the extended theory of planned behavior. *International Journal of Disaster Risk Science, 13*(1), 100–113.

Palen, L., Vieweg, S., Liu, S. B., & Hughes, A. L. (2009). Crisis in a networked world: Features of computer-mediated communication in the April 16, 2007, Virginia Tech event. *Social Science Computer Review, 27*(4), 467–480.

Park, S., McGuinness, K., Fisher, C., Lee, J. Y., et al. (2023). *Digital news report: Australia 2023*. News & Media Research Centre, University of Canberra.

Paton, D. (2019). Disaster risk reduction: Psychological perspectives on preparedness. *Australian Journal of Psychology, 71*(4), 327–341.

Sharp, E. N., & Carter, H. (2020). Examination of how social media can inform the management of volunteers during a flood disaster. *Journal of Flood Risk Management, 13*(4), e12665.

Sheldon, P. (2018). Emergency alert communications on college campuses: Understanding students' perceptions of the severity of a crisis and their intentions to share the alert with parents and friends. *Western Journal of Communication, 82*(1), 100–116.

Simon, T., Goldberg, A., & Adini, B. (2015). Socializing in emergencies—A review of the use of social media in emergency situations. *International Journal of Information Management, 35*(5), 609–619.

Spialek, M. L., & Houston, J. B. (2019). The influence of citizen disaster communication on perceptions of neighbourhood belonging and community resilience. *Journal of Applied Communication Research, 47*(1), 1–23.

Steelman, T. A., & McCaffrey, S. (2012). Best practices in risk and crisis communication: Implications for natural hazards management. *Natural Hazards, 65*, 683–705.

Stephens, K. K., & Robertson, B. W. (2022). Social media platforms and broader participation in crisis communication. In Y. Jin & L. Austin (Eds.), *Social media and crisis communication* (pp. 156–167). Routledge.

Taylor, M., Wells, G., Howell, G., & Raphael, B. (2012). The role of social media as psychological first aid as a support to community resilience building: A Facebook study from 'Cyclone Yasi Update'. *Australian Journal of Emergency Management, 27*(1), 20–26.

Taylor, M., Miller, F., Johnston, K., Lane, A., et al. (2023). *Community experiences of the January–July 2022 floods in New South Wales and Queensland: Final report: Policy-related research findings*. Natural Hazards Research Australia.

Zander, K. K., Sibarani, R., Lassa, J., Nguyen, D., & Dimmock, A. (2022). How do Australians use social media during natural hazards? A survey. *International Journal of Disaster Risk Reduction, 81*, 1–13.

Part V
Conclusion

Chapter 16
Conclusion

Stuart Cunningham, Jane W. Davidson, and Alethea Blackler

Let us start with two pieces of data, one from history and one from economics, and a deduction drawn from sociology and politics. The first, from 85 years ago, occurred when the judge presiding over an Australian Royal Commission into the devastating "Black Saturday" bushfires pronounced "We have not lived long enough". What he meant was that European settlers in this country, Australia, had not learned how to live in a land characterised by climatic extremes of drought, fire and flood. The words echo in environmental historian Tom Griffiths' "we have not yet lived long enough", after his review of the long history of lack of preparedness for such events (2010), despite the repetitiousness with which that lack of preparedness has issued forth from reports and enquiries too numerous to mention here. The second is the Productivity Commission's (2014) finding that 97% of Australia's public funds spent on disasters went to crisis management and recovery and only 3% on preparedness. What we might derive from these points is that "there is no such thing as a natural disaster" (Hartman & Squires, 2006). The history of Indigenous fire management over millennia, leading to early European settlers' puzzlement over what appeared to be curated/estate like landscapes, underscores the fact that human preparedness and the lack of it are material factors in the severity and impact of any "natural" disaster, that is, if we may have lived long enough by *now*.

S. Cunningham (✉)
News & Media Research Centre, University of Canberra, Bruce, ACT, Australia
e-mail: stuart.cunningham@canberra.edu.au

J. W. Davidson
Faculty of Fine Arts and Music, The University of Melbourne, Parkville, VIC, Australia
e-mail: j.davidson@unimelb.edu.au

A. Blackler
School of Design, Queensland University of Technology, Brisbane City, QLD, Australia
e-mail: a.blackler@qut.edu.au

D. Del Favero et al. (eds.), *Climate Disaster Preparedness*, Arts, Research, Innovation and Society, https://doi.org/10.1007/978-3-031-56114-6_16

The introduction to this book sets out its essential framework addressing the acceleration of global warming and the consequent growing unpredictability, nature and scale of fire and flood globally. It notes the growing challenge for models that aim to prepare us for such disasters. It proposes that the technologies of immersive visualisation can be engaged to address this challenge, offering the dynamic ability to "viscerally imagine threats, enact hazardous stories, mock-up risk-laden transactions and probe hyper-local preparedness". The introduction reminds us that this mission integrates and transforms the fragmented arts-versus-science approach, allowing all those potentially facing future cyclones, floods and fires to experience what it could look like and rehearse how to prepare.

The first part, Picturing, reminds us that preparing for climate disasters cannot be conducted under full experimental—what used to be called "laboratory"—conditions. It is far too risky. The take-home message from the modelling science reviewed throughout this section of the book is that we need to develop immersive, adaptive and evolving virtual scenarios that can replicate actual landscapes and disaster dynamics in all their unpredictability. But "alternative" modelling can be very expensive, and there are real limits to its efficacy as it is typically static, two-dimensional modelling. It also requires very specialist work that is hard to do.

The chapters in this first part trace the research developments from 2D and 3D static modelling to visualisation, then to immersive visualisation and then to artificial intelligence-enabled immersive visualisation. The section concludes that the integration of AI with extreme event modelling presents an exciting opportunity for the research community to rapidly develop a deeper understanding of extreme events, as well as the corresponding preparedness, response and management strategies. Looking to the future, the emerging field of "immersive analytics" is explored, emphasising machine learning and artificial intelligence as having the potential to transform disaster modelling with a reduction of demand for extensive data and computation time.

Like Picturing's helpful surveys of research progress being made in the science of extreme event modelling, the book's second part, Narrating, surveys how fields such as the performing arts and screen studies have produced significant work that has not only depicted the threat and the devastation wreaked but that also supported resilience and recovery. This part charts a vast territory in the arts over time, pointing out that surprisingly little has been done in preparedness, despite the trend of growing and now widespread eco-anxiety around extreme weather and its disastrous consequences. It points to crucial societal changes since the 1960s in art, film, television, news and visual effects, charting a shift from "Nat Geo" aesthetic environmentalism to planetary ecosystemics and to mainstreamed awareness of climate change. A key insight is the rise of data visualisation as key to contemporary popular representations of disaster. Included in this part is a discussion of iconic contemporary work in the visualisation of climate change with the *iFire* project, demonstrating the potential of current technologies to be used to rehearse extreme events when the narration and interaction are designed appropriately. This part of the book also underscores the important role that Indigenous knowledges and artistic practices have in understanding human embeddedness within the environment and in thinking about how best to prepare, based on deep knowledge of the nature-culture continuum.

Rehearsing, the book's third part, presents work produced in the fields of architecture, and experience and interaction design, and focuses on how to respond to the complex uncertainties of risk-laden confrontations that disasters present. It emphasises the need for tools and skills development that enable successful practice, rehearsal and improvisation in disaster scenarios and demonstrates how immersive environments can be co-designed with communities to support effective rehearsal of responses to hazardous situations.

This third part presents evidence of the efficacy of approaches such as virtual reality and artificial reality, scenarios, serious games and mock-ups to develop vicarious experience and feed intuition, and their flexibility and adaptability for accommodating diverse scenarios, narratives and means of visualising extreme events. It comes at the question of modelling from artificial intelligence- and machine learning-enhanced architectural design, putting forward two research approaches—space syntax and intelligent mobility modelling—in advancing the research field that this book covers. It is clear about the limits of these approaches, particularly their dependence on known or predictable properties in land- and streetscapes. Nevertheless, this part, borrowing the language of the performing arts, stresses the usefulness of such models "when confronted with unpredictable events, these examples treat people's behaviours as needing practice, rehearsal and improvisation".

While Rehearsing responds to the book's pervasive theme of uncertainty, Communicating, the book's fourth part, starts with a discussion of the policy importance of "preparedness under uncertainty". In Picturing, we are reminded that "one of the major open questions in visualisation approaches is how to usefully communicate uncertainty in predictions and forecasts". In Rehearsing, we are reminded that wildfires and floods alter the landscape, blocking roads, destroying landmarks and turning the built environment and infrastructure into potential hazards, which renders real-time movement disorientating and uncertain. These types of uncertainties form a challenge that we can take on using methods developed for narrating and rehearsing extreme events, such as developing and visualising realistic scenarios, or, for wider exploration of alternative action and outcomes, employing dynamic interactive narrative approaches. The ability for a wide range of community members and stakeholders as well as responders to understand unpredictability; to realistically explore alternative outcomes based on flood or fire behaviour, altered environments or various responses by responders and actors in the scenario; and to be well trained for communicating under such conditions will be essential to developing true community preparedness.

Communicating focuses on communication in disaster scenarios and critiques current practices focused on "top-down" approaches during and in the aftermath of disaster events and governments' reliance on established communities of practice such as emergency workforces. It reflects on official independent reports into flood and fire disasters in the last five years in Australia, highlighting dangerous gaps in official policy responses—a distinct lack of focus on preparedness and an almost complete lack of focus on the role that arts, culture and creativity can play in dealing with climate emergencies.

The book's focus is on what specific disciplines can provide in the way of representational resources to better prepare individuals, communities and populations to deal with unpredictable but worsening climate scenarios. It recurs in Communicating and, with it, the additional challenge *of* how disciplines must transform and collaborate. This is exemplified in its survey of the communication literature showing the extent to which social capital can be built in community use of social media during emergencies. The challenge, however, is to understand how such social capital, once built up in crisis, can continue to evolve post-crisis.

16.1 Final Words

This volume has shown that a cross-disciplinary approach to climate disaster preparedness, involving socially embedded creative arts practice melded with advanced technology, represents real and needed research innovation. The case is advanced for the use of arts and cultural knowledge, including that of Indigenous communities, in concert with experimental immersive visualisation technologies to shift the formulation from static observation to stakeholders viscerally interacting with unforeseen threat scenarios. Such a forward research agenda can deliver a form of sensorial plausibility enabled by the arts as well as a physical probability technology that science can deliver. Presenting this agenda, the authors open possibilities for research, industry, emergency and resident communities to collaborate on visualisations embedded in personal experiences, modelling and creative imagination to ensure that we can prepare better for the impending challenges ahead.

References

Griffiths, T. (2010). We have still not lived long enough: Black Friday and Black Saturday, *Humanities Australia, 1*, 23–33.

Hartman, C., & Squires, G. (Eds.). (2006). *There is no such thing as a natural disaster: Race, class, and Hurricane Katrina*. Routledge.

Productivity Commission. (2014). *Natural disaster funding arrangements. Inquiry report no. 74*, Canberra: JEL code: H77, H84

MIX
Papier aus verantwortungsvollen Quellen
Paper from responsible sources
FSC® C105338

If you have any concerns about our products,
you can contact us on
ProductSafety@springernature.com

In case Publisher is established outside the EU,
the EU authorized representative is:
Springer Nature Customer Service Center GmbH
Europaplatz 3, 69115 Heidelberg, Germany

Printed by Libri Plureos GmbH
in Hamburg, Germany